Fluorescence and Phosphorescence Spectroscopy

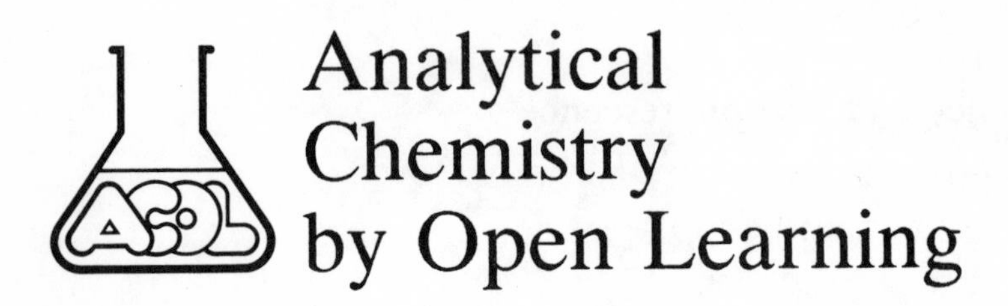
Analytical
Chemistry
by Open Learning

Titles in Series:

Samples and Standards
Sample Pretreatment and Separation
Classical Methods
Measurement, Statistics and Computation
Using Literature
Instrumentation
Chromatographic Separations
Gas Chromatography
High Performance Liquid Chromatography
Electrophoresis
Thin Layer Chromatography
Visible and Ultraviolet Spectroscopy
Fluorescence and Phosphorescence Spectroscopy
Infra Red Spectroscopy
Atomic Absorption and Emission Spectroscopy
Nuclear Magnetic Resonance Spectroscopy
X-ray Methods
Mass Spectrometry
Scanning Electron Microscopy and X-Ray Microanalysis
Principles of Electroanalytical Methods
Potentiometry and Ion Selective Electrodes
Polarography and Other Voltammetric Methods
Radiochemical Methods
Clinical Specimens
Diagnostic Enzymology
Quantitative Bioassay
Assessment and Control of Biochemical Methods
Thermal Methods
Microprocessor Applications

Fluorescence and Phosphorescence Spectroscopy

Analytical Chemistry by Open Learning

Author:
DAVID RENDELL
Wolverhampton Polytechnic

Editor:
DAVID MOWTHORPE

on behalf of ACOL

Published on behalf of ACOL, Thames Polytechnic, London
by
JOHN WILEY & SONS
Chichester · New York · Brisbane · Toronto · Singapore

Library of Congress Cataloging in Publication Data:

Rendell, David.
Fluorescence and phosphorescence.
(Analytical Chemistry by Open Learning)
1. Fluorescence spectroscopy—Programmed instruction.
2. Fluorimetry—Programmed instruction.
3. Phosphorimetry—Programmed instruction.
4. Phosphorescence spectroscopy—Programmed instruction.
5. Chemistry, Analytic—Programmed instruction.
I. Mowthorpe, David. II. ACOL (Project)
III. Title IV. Series: Analytical Chemistry by Open Learning (Series).
QD96.F56R46 1987 543'.0858 87–8157
ISBN 0 471 91380 4

ISBN 0 471 91381 2 (pbk.)

British Library Cataloguing in Publication Data:

Rendell, David.
Fluorescence and phosphorescence
(Analytical chemistry).
1. Fluorescence 2. Phosphorescence
I. Title II. Mowthorpe, David J.
III. ACOL IV. Series
535'.35 QC477
ISBN 0 471 91380 4
ISBN 0 471 91381 2 Pbk

Printed and bound in Great Britain

Analytical Chemistry

This series of texts is a result of an initiative by the Committee of Heads of Polytechnic Chemistry Departments in the United Kingdom. A project team based at Thames Polytechnic using funds available from the Manpower Services Commission 'Open Tech' Project has organised and managed the development of the material suitable for use by 'Distance Learners'. The contents of the various units have been identified, planned and written almost exclusively by groups of polytechnic staff, who are both expert in the subject area and are currently teaching in analytical chemistry.

The texts are for those interested in the basics of analytical chemistry and instrumental techniques who wish to study in a more flexible way than traditional institute attendance or to augment such attendance. A series of these units may be used by those undertaking courses leading to BTEC (levels IV and V), Royal Society of Chemistry (Certificates of Applied Chemistry) or other qualifications. The level is thus that of Senior Technician.

It is emphasised however that whilst the theoretical aspects of analytical chemistry can be studied in this way there is no substitute for the laboratory to learn the associated practical skills. In the U.K. there are nominated Polytechnics, Colleges and other Institutions who offer tutorial and practical support to achieve the practical objectives identified within each text. It is expected that many institutions worldwide will also provide such support.

The project will continue at Thames Polytechnic to support these 'Open Learning Texts', to continually refresh and update the material and to extend its coverage.

Further information about nominated support centres, the material or open learning techniques may be obtained from the project office at Thames Polytechnic, ACOL, Wellington St., Woolwich, London, SE18 6PF.

How to Use an Open Learning Text

Open learning texts are designed as a convenient and flexible way of studying for people who, for a variety of reasons cannot use conventional education courses. You will learn from this text the principles of one subject in Analytical Chemistry, but only by putting this knowledge into practice, under professional supervision, will you gain a full understanding of the analytical techniques described.

To achieve the full benefit from an open learning text you need to plan your place and time of study.

- Find the most suitable place to study where you can work without disturbance.
- If you have a tutor supervising your study discuss with him, or her, the date by which you should have completed this text.
- Some people study perfectly well in irregular bursts, however most students find that setting aside a certain number of hours each day is the most satisfactory method. It is for you to decide which pattern of study suits you best.
- If you decide to study for several hours at once, take short breaks of five or ten minutes every half hour or so. You will find that this method maintains a higher overall level of concentration.

Before you begin a detailed reading of the text, familiarise yourself with the general layout of the material. Have a look at the course contents list at the front of the book and flip through the pages to get a general impression of the way the subject is dealt with. You will find that there is space on the pages to make comments alongside the text as you study—your own notes for highlighting points that you feel are particularly important. Indicate in the margin the points you would like to discuss further with a tutor or fellow student. When you come to revise, these personal study notes will be very useful.

Π When you find a paragraph in the text marked with a symbol such as is shown here, this is where you get involved. At this point you are directed to do things: draw graphs, answer questions, perform calculations, etc. Do make an attempt at these activities. If necessary cover the succeeding response with a piece of paper until you are ready to read on. This is an opportunity for you to learn by participating in the subject and although the text continues by discussing your response, there is no better way to learn than by working things out for yourself.

We have introduced self assessment questions (SAQ) at appropriate places in the text. These SAQs provide for you a way of finding out if you understand what you have just been studying. There is space on the page for your answer and for any comments you want to add after reading the author's response. You will find the author's response to each SAQ at the end of the text. Compare what you have written with the response provided and read the discussion and advice.

At intervals in the text you will find a Summary and List of Objectives. The Summary will emphasise the important points covered by the material you have just read and the Objectives will give you a checklist of tasks you should then be able to achieve.

You can revise the Unit, perhaps for a formal examination, by re-reading the Summary and the Objectives, and by working through some of the SAQs. This should quickly alert you to areas of the text that need further study.

At the end of the book you will find for reference lists of commonly used scientific symbols and values, units of measurement and also a periodic table.

Contents

Study Guide

Photoluminescence, which comprises the techniques of fluorescence and phosphorescence, is one of the major 'growth areas' in analytical chemistry today. Although fluorescence was one of the earliest instrumental technique available to the analyst, recent developments in instrumentation and sample handling techniques have only now made it possible for the full potential, particularly its very high sensitivity, to be realised in everyday analysis.

This Unit provides an introduction to fluorescence and phosphorescence techniques to meet the needs both of students seeking a general education in analytical chemistry and of established analysts who have recently become aware of the technique and want to find out more about them and what they have to offer.

Although the theoretical basis of all spectroscopic techniques is somewhat complicated, the simple treatment of the basic principles of photoluminescence given in Part 1 should provide an adequate background for the understanding of its analytical applications. This should be readily intelligible if you have studied chemistry to A level though further study, perhaps by the way of an HNC course would be an advantage. You should certainly make sure that you are clear about the nature of electromagnetic radiation and the wavelength range of the visible and ultra-violet regions. It will also help if you are aware of the processes associated with the absorption and emission of radiation and the quantitative aspects of absorption as expressed by the Beer–Lambert Law, though the opportunity to revise these topics will be given as they arise in the Unit.

The instrumentation required for photoluminescence work is described in Part 2 in simple terms using block diagrams avoiding undue detail about individual components. However, if you have some appreciation of elementary optics you will find that useful.

Quantitative aspects of the techniques are presented in Part 3. Comparisons are made with absorption techniques with which you will probably be familiar, particularly if you have had some experience of work in an analytical laboratory. Familiarity with the routine operations and calculations of analytical chemistry will be assumed but again plenty of practice will be provided and the answers to questions will provide help, if needed.

The remainder of the Unit is devoted to a survey of the methods available for both organic and inorganic analytes using natural fluorescence or the formation of fluorescent derivatives. This will include separate sections on some of the more exciting developments such as fluoroimmunoassay and room temperature phosphorescence. The Unit concludes with a group of Case Studies which are particularly important in providing some of the experience normally gained through practical work which will of necessity, be somewhat restricted in a distance learning course.

Supporting Practical Work

Analytical chemistry is essentially a practical subject and you cannot consider yourself to have a full appreciation of any analytical technique until you have tried it out in the laboratory for yourself. For this Unit you will need to find an institution equipped with a spectrofluorimeter of some kind or at least a fluorescence attachment to a uv/visible spectrophotometer. With this instrument you should seek to attain the following objectives:

(*a*) to observe the fluorescence of quinine (or any suitable fluorescent compound) and to determine the wavelengths at which the excitation and emission are at a maximum;

(*b*) to record the excitation and emission spectra of quinine and/or one or two other typical fluorescent compounds;

(*c*) to plot the intensity/concentration curve for quinine (or other compound) over a wide range of concentration to demonstrate non-linear behaviour and the inner filter effect;

(*d*) to determine the concentration of quinine in tonic water using the linear section of the intensity/concentration curve as the calibration curve;

(*e*) to demonstrate the effect of quenching by halide ions and other species on the fluorescence of quinine.

(*f*) to determine the concentration of a metal or an organic compound at trace levels using a fluorimetric reagent or a derivatisation procedure.

Objectives (*a*) and (*b*) can only be fully realised with a dual monochromator spectrofluorimeter though a limited study could be carried out with a filter/monochromator instrument.

Objectives (*c*) to (*f*) require only a simple filter fluorimeter providing the appropriate filters are available.

The full list of objectives should be achieved in about six hours of laboratory work with technical assistance to provide the necessary primary standard solution and reagents.

If you are fortunate enough to get your hands on an instrument with the capability of recording phosphorescence, the objectives would be extended to include demonstrations of room temperature phosphorescence with samples such as quinoline in β-cyclodextrin or a phosphorescence derivative on filter paper and long-lived luminescence with a complex of europium. However the only essential exercises for this Unit are those involving fluorescence.

Bibliography

Bibliography

1. 'STANDARD' ANALYTICAL CHEMISTRY TEXTBOOKS

Nearly all the books giving a comprehensive treatment of instrumental analytical chemistry contain a chapter on fluorescence spectroscopy.

Appropriate examples are:

(*a*) D. A. Skoog, *Principles of Instrumental Analysis*, Saunders College Publishing, 3rd edition 1985.

(*b*) H. H. Willard, L. L. Merritt, J. A. Dean and F. A. Settle, *Instrumental Methods of Analysis*, Van Nostrand, 6th edition, 1981.

2. SPECIALIST TEXTBOOKS

(*a*) C. A. Parker, *Photoluminescence of Solutions*, Elsevier, 1968.

(*b*) S. G. Schulman, *Molecular Luminescence Spectroscopy, Methods and Applications*, Wiley 1985.

(*c*) S. G. Schulman and D. V. Naik, *Fluorescence and Phosphorescence Spectroscopy: Physicochemical Principles and Practice*, Pergamon Press, 1977.

(*d*) E. L. Wehry (editor), *Modern Fluorescence Spectroscopy*, Plenum Press, Vol 1 and 2 1976, Vol 3 and 4 1981.

Acknowledgements

Figures 1.2c, 4.2b and 5.4a are redrawn from C. A. Parker, *Photoluminescence of Solutions*, Elsevier, 1968, with the permission of the Elsevier Publishing Co.

Figure 1.4a is based on J. A. Barltrop and J. D. Coyle, *Excited States in Organic Chemistry*, Figure 3.2, John Wiley and Sons, 1975.

Figure 2.2a is reproduced by courtesy of Perkin Elmer, Beaconsfield.

Figures 2.7b and 2.7c are redrawn from T. C. O'Haver and J. D. Winefordner, *Analytical Chemistry*, **38**, 602, 1966, with permission of the American Chemical Society.

Figures 5.5b and 5.5c are redrawn from R. A. Chalmers and G. A. Wadds, *Analyst*, **95**, 234, 1970 with permission of the Royal Society of Chemistry.

The Figures in Section 5.8 are redrawn from S. Scypinski and L. J. Cline Love, *Analytical Chemistry*, **56**, 322, 1984 with permission of the American Chemical Society.

1. Introduction

1.1. THE NATURE OF LUMINESCENCE

The production of light is most commonly associated with a process involving heat. Many common sources of radiation rely on the principle that when a material is heated it begins to glow, first becoming red, then orange and finally white as the temperature is raised. This is the basis of incandescent light sources such as the electric filament lamp and the gas mantle.

Π The heat of a flame or the electrical energy in a discharge can also be used to produce light from metal atoms, a familiar example of the latter being the discharge lamps used for street lighting. Do you happen to know what metals they contain?

The yellow lamps contain sodium and the blue–green lamps contain mercury.

There are, however, cases in which the emission of light is not associated with thermal or electrical energy. Such 'cold light' phenomena as they are called are much less common but include the light emitted by glow-worms and fireflies, 'light sticks' which are tubes which give out light when the chemicals they contain are mixed, luminous paint and theatrical effects in which light is produced from special dyes which glow when irradiated by ultraviolet radiation on an otherwise dark stage. These phenomena are all forms of 'luminescence' which is the general term to describe processes leading to the emission of light.

The emission of light is always a response to an input of energy of some type and the different types of luminescence are distinguished by the addition of a prefix to denote the type of energy involved. Thus the light from a glow-worm is the result of a biological process and is therefore called 'bioluminescence'. Light released as a result of a chemical reaction is called 'chemiluminescence'. Luminous paint contains a compound which glows when radioactive particles, usually provided by tritium (^{3}H) atoms, interact with it and is therefore 'radioluminescent'. Finally, light produced by illuminating a compound with uv radiation receives its energy from the incident uv photons and so is called 'photoluminescence'.

Although all forms of luminescence are potentially useful in analysis, photoluminescence is by far the most important and this is the subject of the present unit.

SAQ 1.1a

Select a source of light from the list below which is an example of

(*A*) radioluminescence

(*B*) photoluminescence

(*C*) bioluminescence

(*D*) chemiluminescence

Sources:

(*i*) cold light sticks.

(*ii*) a theatrical mask which glows on a darkened stage. ⟶

SAQ 1.1a (cont.)

(*iii*) glow-worms.

(*iv*) luminous paint.

1.2. THE FLUORESCENCE MECHANISM

At present, the most widely used type of photoluminescence in analytical chemistry is fluorescence. Fluorescence was first observed as long ago as 1565 and gets its name from the fact that the mineral fluorspar was found to glow under uv radiation. In all forms of luminescence it is necessary for a molecule to be raised into an excited electronic state before it can emit radiation. Fluorescence is distinguished from other types of photoluminescence by the fact that the excited molecule returns to the ground state immediately after excitation, the time spent in the excited state being typically of the order of 10^{-8} s. This 'lifetime' of the excited state, though apparently very short, is in fact very long compared with other events on the molecular scale, some of which are listed below.

Time for transitions between electronic states	10^{-15} s
Time period for molecular vibrations	10^{-14} s
Time period for molecular rotations	10^{-11} s
Average time between collisions for a molecule in the liquid phase at room temperature	10^{-12} s

Π Use the data given above to calculate

(*a*) the number of vibrations,
(*b*) the number of rotations and
(*c*) the number of collisions

which occur with an electronically excited molecule before it returns to the ground state with the emission of fluorescence.

You should get the following answers:

(*a*) 1,000,000. If the molecule remains in the excited state for 10^{-8} s and one vibration takes 10^{-14} s, the number of vibrations occurring during the lifetime of the excited molecule is $10^{-8}/10^{-14} = 10^{6}$.

(*b*) 1,000. (The calculation is similar to that in (*a*))

(*c*) 10,000.

Before we get to grips with the fluorescence mechanism, let's be quite sure that we are clear about the absorption process. When a molecule absorbs a photon of uv radiation, it undergoes a transition to an excited electronic state and one of its electrons is promoted to an orbital of higher energy. The transition is classified by reference to the orbitals concerned.

Π There are two important types of transition for organic molecules which you have probably met before. Can you remember what they are called?

$n \rightarrow \pi^*$ and $\pi \rightarrow \pi^*$.

Π Identify the electrons involved in each of these types of transition in the case of propanone (acetone), CH_3COCH_3.

$n \rightarrow \pi^*$: one of the lone pair electrons on the oxygen atom which is in a non-bonding orbital.

$\pi \rightarrow \pi^*$: one of the pair of electrons in the π bonding orbital of the C=O group.

In both cases the electron finishes up in a anti-bonding orbital, not necessarily the same one – there are plenty to choose from! (Even if the electrons do end up in the same orbital, the energies of the excited states produced are different because the electronic configurations are different – there is a 'gap' in a different lower energy orbital.)

The $n \rightarrow \pi^*$ transition is of lower energy in simple molecules like propanone but in more complex molecules like aromatic ketones, where there are several π bonding as well as anti-bonding orbitals, this is not always the case.

The bands in the uv spectrum corresponding to these transition are also called $n \rightarrow \pi^*$ and $\pi \rightarrow \pi^*$ bands. These differ in two important respects. Do you know what they are?

(*a*) The $n \rightarrow \pi^*$ band is usually at longer wavelength (since the transition energy is lower).

(*b*) The $n \rightarrow \pi^*$ band is always much weaker.

In the case of propanone, the $n \rightarrow \pi^*$ band is at 270 nm and has a molar absorptivity of 20 $dm^3\ mol^{-1}\ cm^{-1}$. The $\pi \rightarrow \pi^*$ band is at 180 nm and has a molar absorptivity of about 1000 $dm^3\ mol^{-1}\ cm^{-1}$. The molar absorptivity, ϵ, is defined by the Beer–Lambert Law which you have probably already come across.

Π Can you remember how this law is usually expressed?

The most common form is $A = \epsilon c d$ where A is the absorbance of a solution of concentration c mol dm^{-3} in a cell of path-length d cm. (The alternative units for molar absorptivity are $m^2\ mol^{-1}$. If we express our values above in these units they become 2 and 100 $m^2 mol^{-1}$.)

The electronic energy is not the only type of energy affected when a molecule absorbs a photon of uv radiation. Molecules also possess vibrational energy which is quantised to give a series of vibrational energy levels. Organic molecules have a large number of vibrations. Each one of these contributes a series of almost equally-spaced vibrational levels to each electronic state.

Let's now follow a molecule through the entire process which results in the emission of fluorescence. Initially, in common with virtually all other molecules in the sample, it will be in its electronic ground state.

Π What will be the most likely vibrational state of the molecule at room temperature?

Unless the molecule possesses some very low frequency vibrations, at least 99% of the molecules in the sample will be in the lowest level of every vibration.

In order to simplify the discussion, let's assume that only one of the vibrations gets excited and that when a molecule absorbs a photon of uv radiation the resulting transition starts from the lowest level of this vibration in the electronic ground state.

After the absorption the molecule will be in an excited electronic state but now, depending on the energy of the photon absorbed, the vibration may also be excited. In order to identify the specific vibrational levels involved in the transitions between electronic states, spectroscopists label them with their vibrational quantum number, v. If the level is in the excited electronic state this is indicated by a single dash (v′) while if it is in the ground electronic state a double dash v″) is used.

Fig. 1.2a shows a simplified energy level diagram on which the vibrational levels are labelled with their v′ or v″ values. We can now show the transitions corresponding to the absorption of radiation by upward-pointing arrows as on the left-hand side of the diagram and refer to any particular transition by quoting the v values in the order v″,v′ (lower state first). For example, the transition on the far left of Fig. 1.2a is the (0,0) transition.

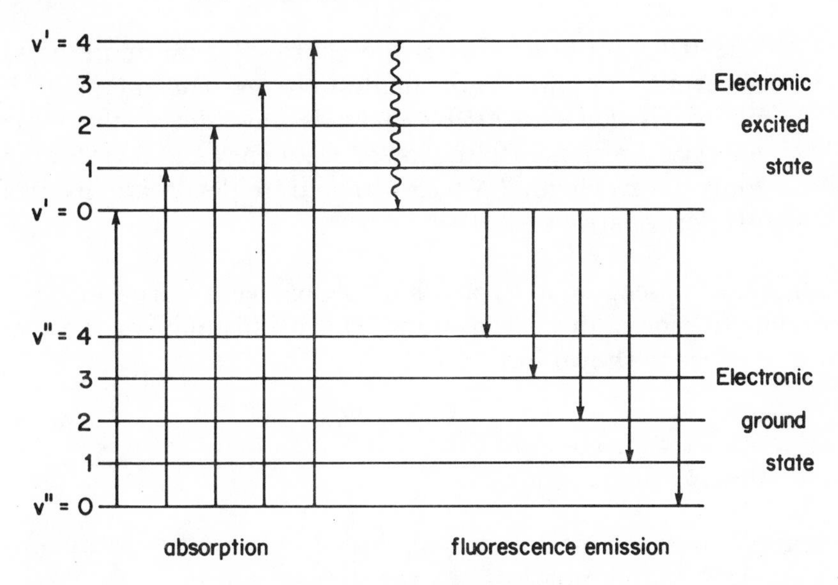

Fig. 1.2a. *A simple energy level diagram showing molecular electronic and vibrational energy levels*

If you have studied other spectroscopic units, you will be familiar with energy level diagrams of this type which show the various possible quantised energy values available to a molecule on a vertical scale of increasing energy. If not, you will find this type of presentation helpful in discussing the origin of features in spectra and you should take note of it because it will crop up again quite frequently in this Unit. If you are the kind of person who enjoys name-dropping you can call it a 'Jablonski diagram' after the man who first devised it!

Π What is the relationship between the lengths of the arrows in Fig. 1.2a and the energy of the photon absorbed?

The energy of the photon is proportional to the length of the arrow. The vertical scale of Fig. 1.2a is linear in energy and the length of the arrows represents a difference in energy between two energy levels. This is equal to the energy of the photon involved in the transition.

Π Which of the transitions in Fig. 1.2a corresponds to the absorption of radiation of longest wavelength?

The 0,0 transition. This is the shortest arrow representing absorption of photons of smallest energy. The energy of the photon is equal to hν by Planck's Law and so the lowest energy corresponds to the lowest frequency and therefore the longest wavelength. (Wavelength and frequency are inversely proportional)

With some molecules it is possible to observe absorption corresponding to each of these transitions. This is illustrated by the spectrum of benzene shown in Fig. 1.2b.

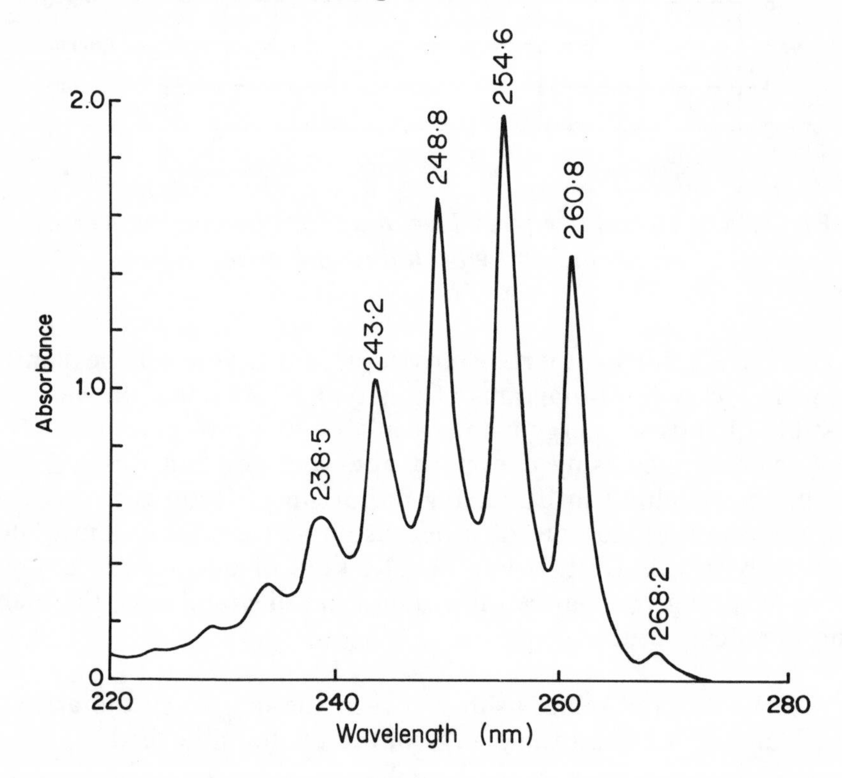

Fig. 1.2b. *Ultraviolet absorption spectrum (220 to 280 nm) of benzene*

The absorption in this spectrum is made up of a number of sharp features to give an overall electronic absorption band which shows 'vibrational fine structure'. The intensities of the individual peaks in the fine structure is governed by the 'probability' of the transition, the highest intensity corresponding to the most probable transition.

(This principle arises out of the quantum theory, the details of which do not concern us in this course. It is as well however to be aware of the terminology since you may come across it in the literature of analytical chemistry.)

This spectrum is a little unusual since the uv spectra of most liquid samples consist of broad bands having a smooth contour and little, if any, fine structure. This loss of structure is due to a number of factors, chief of which is the interaction between molecules in the liquid phase. This causes a small shift in the energy levels of individual molecules, particularly if the molecule is polar or dissolved in a polar solvent, so that the transitions in different molecules are of slightly different energy and the absorptions overlap. A good example of this behaviour is the spectrum of a solution of phenol in hexane which shows only very poorly resolved fine structure. When a polar solvent such as ethanol is used, the fine structure disappears completely to leave a band with a rather irregular profile. Even so the wavelength of maximum intensity still corresponds to the most probable transition.

Π The previous discussion indicates that phenol is more polar than benzene and ethanol is more polar than hexane. Can you explain why this is so?

You will need to be aware of the structural formulae of these compounds to answer this question! These are shown below.

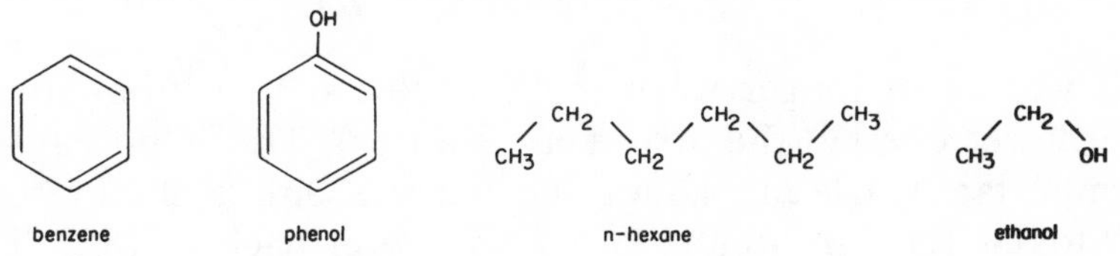

OH

CH_3 CH_2 CH_2 CH_2 CH_2 CH_3

CH_3 CH_2 OH

benzene phenol n-hexane ethanol

Benzene and hexane are both hydrocarbons in which all the bonds are either between a carbon atom and a hydrogen atom or between two carbon atoms. In these bonds the electrons are equally shared between the atoms and no separation of charge occurs. Phenol and ethanol both contain hydroxyl groups (OH) in which the oxygen atom attracts electrons away from the carbon and hydrogen atoms because of its greater 'electronegativity'. This results in the oxygen becoming negatively charged and the carbon and hydrogen positively charged. This separation of charge gives rise to a quantity called the 'dipole moment' which is defined as the product of the charge and the distance between the positive and negative centres. Molecules with a dipole moment are said to be 'polar'. The dipole moment is a very important quantity in determining the intensities of bands in optical (uv/visible, ir and microwave) spectra.

The excess vibrational energy of the excited molecules is rapidly dissipated through thermal motion (collisions) and other processes which occur much more rapidly than the direct emission of a photon from the excited state. Consequently, within a very short time (less than 10^{-10} s) an excited molecule will 'drop' to its lowest vibrational level, $v' = 0$. This process is called *vibrational relaxation* and is represented by the curly arrow in the middle of Fig. 1.2a. When fluorescence emission does occur therefore, all the transitions start from the $v' = 0$ level but they may terminate in any of the vibrational levels of the electronic ground state as shown by the arrows on the right-hand side of Fig. 1.2a. These transitions give rise to the fluorescence emission band of the molecule which, like the absorption band, may, but more often does not, show fine structure. Again the most intense region of the band corresponds to the most probable transition(s).

Π Would you expect the fluorescence band to be of longer or shorter wavelength than the absorption band?

It will be of longer wavelength since the transitions involved are of lower energy (the arrows are shorter). This is perhaps the most important single fact about fluorescence, and is often referred to as 'Stokes' Law' in honour of its discoverer back in 1852. This property of effectively being able to change the wavelength of radiation is used commercially to produce 'whiter-than-white' detergents. By incorporating a colourless fluorescent compound into a washing powder, traces remain in the fabric after washing. The fabric then absorbs uv radiation and emits at the blue end of the visible spectrum thus apparently reflecting more light than is falling on it.

If we plot the absorption and emission spectra of a compound on the same chart, the displacement of the emission band to longer wavelength is immediately apparent. This is illustrated by the spectra of anthracene in Fig. 1.2c.

Π There is another feature of the spectra that is brought out by plotting them together and which is particularly noticeable with anthracene because of the marked fine structure. See if you can spot what this is.

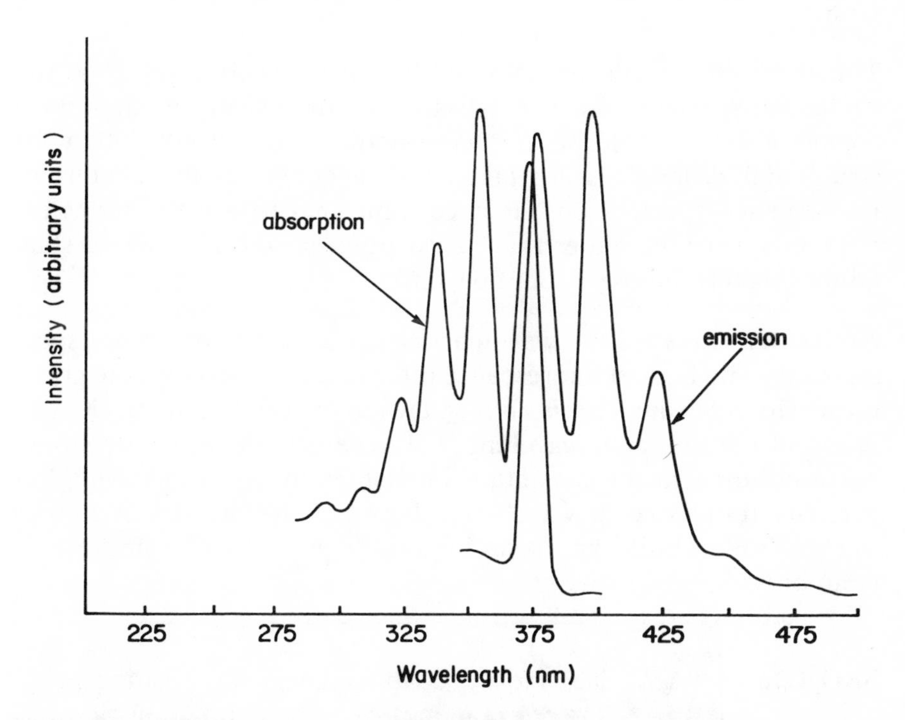

Fig. 1.2c. *Absorption and emission (fluorescence) spectra of anthracene*

The emission and absorption spectra bear a 'mirror-image' relationship to each other. Draw a vertical line on Fig. 1.2c at a wavelength of 375 nm. This divides the diagram into two almost symmetrical halves.

The more sophisticated instruments used in fluorescence spectroscopy are in fact capable of recording both spectra. The emission spectrum is obtained by irradiating the sample at the wavelength of maximum absorption and observing the emitted fluorescence with a scanning monochromator which gives a plot of intensity versus wavelength. The absorption spectrum is obtained by plotting the intensity of the fluorescence emission as the wavelength of the exciting radiation is changed.

(The instrumentation will be discussed in detail in Part 2 of this Unit.)

The absorption spectrum obtained in this way is referred to as the *excitation spectrum* but it is identical to the spectrum run with a conventional uv absorption spectrophotometer provided that instrumental differences are taken into account. In the context of fluorescence spectroscopy, it is conventional to use the term 'excitation spectrum' rather than 'absorption spectrum' and we shall adopt this terminology from now on.

Notice that there are two wavelengths involved in fluorescence spectroscopy. The highest fluorescence intensity, and therefore the greatest sensitivity in analysis, is obtained when the excitation wavelength, λ_{ex}, and the emission wavelength, λ_{em}, are both set to the values corresponding to the maximum intensities in the respective spectra. The difference in wavelength between the maxima is known as the 'Stokes' Shift' and is an important property of a fluorescent compound.

SAQ 1.2a

(*i*) The most probable electronic transition in the benzene molecule is the 0,1 transition. What is the wavelength of the absorption corresponding to this transition? Refer to Fig. 1.2b. ⟶

SAQ 1.2a (cont.)

(*ii*) What are the wavelengths of the 0,0 and the 0,3 bands?

(*iii*) Which of these bands corresponds to the smallest transition energy?

(N.B. The weak band at 268.2 nm involves one of the other vibrations of benzene which we have ignored in our discussion.)

There is some ambiguity in the use of the word 'band' in discussing uv spectra. The term '0,1 band' refers to one specific feature in the fine structure of the overall 'electronic band'. The meaning is generally clear from the context in which it is used.

SAQ 1.2b

Choose words from the following list to fill the blanks in the paragraph below:

longer shorter higher lower

(each word may be used once, twice, or not at all)

'The most important feature of the transitions giving rise to the fluorescence band is that they are of __________ energy than those associated with the absorption band and so the arrows are __________ on the energy scale of Fig. 1.2a. Consequently, the fluorescence emission band appears at __________ frequency and therefore at __________ wavelength than the absorption band'.

SAQ 1.2c

Draw a simple energy level diagram to show the ground and first excited electronic states of a molecule with *five* vibrational levels in each state.

(*i*) Label the vibrational levels with the quantum number v′ or v″.

(*ii*) Identify the level in which most molecules will be found at room temperature.

(*iii*) Identify the level from which fluorescence emission is most likely to originate.

(*iv*) Draw an arrow to show the transition giving rise to the 0,4 absorption band.

(*v*) Draw an arrow to show the transition giving rise to the 0,1 emission band.

(*vi*) Which band, (*iv*) or (*v*), has the longer wavelength?

SAQ 1.2d

Which transition gives rise to a band in both the excitation and emission spectrum of a fluorescent compound?

SAQ 1.2e

Evaluate the Stokes' Shift for anthracene (refer to Fig. 1.2c).

1.3. SOLVENT EFFECTS IN FLUORESCENCE SPECTROSCOPY

Although the 0,0 transition gives rise to a band in both the fluorescence excitation and emission spectra, its wavelength may not in fact be identical in both spectra. A small displacement to longer wavelength is generally observed in the emission spectrum due to the interaction between the fluorescent solute molecules and molecules of the solvent. We have already seen that the polarity of the solvent has a fairly profound effect upon the the appearance of fine structure in uv absorption bands so it is not surprising to find that this same property is the major factor in causing the shift in the 0,0 band.

It is a general point of some importance in analytical spectroscopy to note that the solvent can have a significant effect upon the appearance of a spectrum. You should always bear this in mind when comparing spectra recorded with the sample in different solvents – a frequent occurrence when the spectra originate from different laboratories. The underlying reason for these effects is perhaps of less direct analytical significance but a deeper understanding of the processes involved may save you from getting caught out in an analytical investigation!

In a solution the solvent molecules distribute themselves around the solute molecules in a way which leads to the greatest stability for the system. This is a fundamental law of nature which is expressed thermodynamically in terms of the energy of the system.

Π In the most stable configuration, would you expect the energy of the system to be at a maximum or a minimum?

The greatest stability corresponds to the minimum energy.

The energy of a solute/solvent system is determined to a large extent by electrostatic forces and so the interaction is greatest when we have a polar solute in a polar solvent. This leads to the greatest difference between the most stable and least stable configurations. When a molecule undergoes electronic excitation there is usually a significant change in its polarity. Consequently, the distribution of solvent molecules which gave the most stable configuration for the ground state does not necessarily minimise the energy for the same molecule in the excited state. The absorption process is virtually instantaneous compared with molecular motion and the solvent molecules subsequently adjust to achieve the most stable configuration for the excited state with a slight reduction in the energy. It is from this state that the molecule returns to the ground state where again the energy of the system is above the minimum because the distribution of the solvent molecules corresponds to the excited state situation. Subsequent adjustment then reduces the energy to the value from which the excitation originally occurred and the ground state environment is restored. Consequently the emitted photon is of lower energy than the exciting photon and the 0,0 band appears at longer wavelength in the emission spectrum. The entire process is illustrated in Fig. 1.3a.

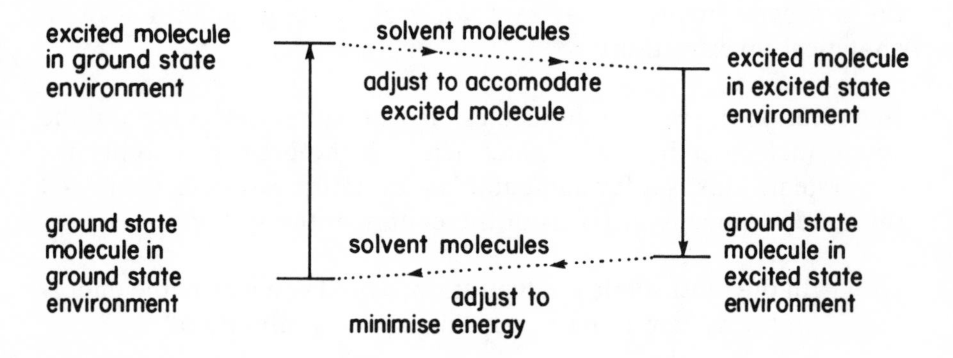

Fig. 1.3a. *Energy changes due to spectroscopic transitions and solute-solvent interactions*

SAQ 1.3a

Which of the following times would you consider to be reasonable to allow adjustment of the solvent molecules in the sequence of Fig. 1.3a to take place:

10^{-20} s, 10^{-15} s, 10^{-10} s, 10^{-5} s ?

SAQ 1.3b

In the diagram below, the central shaded ellipse represents an excited molecule with its polarity indicated by + – at the centres of electrical charge. The surrounding unshaded ellipses represent polar solvent molecules.

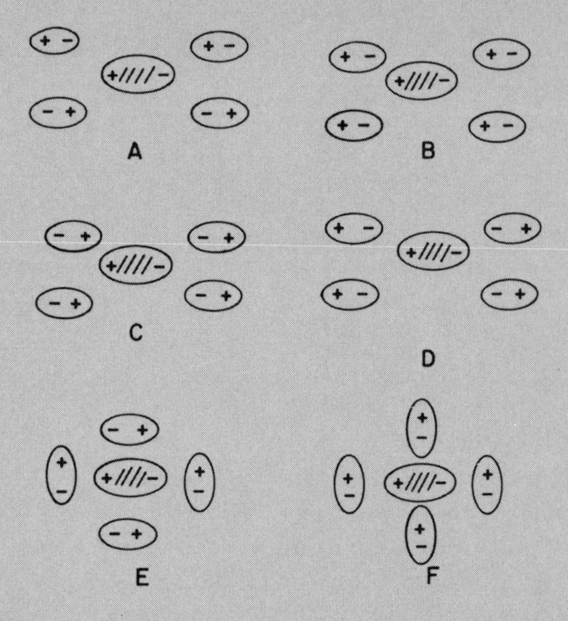

(*i*) Which of the configurations A to F is likely to be of lowest energy?

(*ii*) Which of the configurations of solvent molecules would be the most stable for the same solute molecule in the ground state if this were:

— of opposite polarity to the excited state;
— non-polar?

SAQ 1.3b

1.4. FLUORESCENCE AND MOLECULAR STRUCTURE

You should by now have a reasonably clear idea of the process leading to the emission of fluorescence, the time scale of the steps in this process and the influence of the solvent on the appearance of the spectrum. In practice it is found that only a very small proportion of organic compounds fluoresce and we shall now attempt to identify common features in the structure of the molecules of those that do.

Before a molecule can emit radiation by the fluorescence process it must of course be able to absorb radiation, but by no means all molecules which absorb uv or visible radiation are fluorescent.

As a first step, it will be useful to establish a method of quantifying the extent to which a given molecule fluoresces. This is done by means of the *quantum yield* or *fluorescence efficiency*, ϕ_f, which is defined as the fraction of the incident radiation which is re-emitted as fluorescence.

$$\phi_f = \frac{\text{no. of photons emitted}}{\text{no. of photons absorbed}} = \frac{\text{intensity of fluorescence}}{\text{intensity of absorption}}$$

Values of ϕ_f fall in the range 0 to 1 and are an inherent property of a molecule determined to a large extent by its structure. A high value of ϕ_f is generally associated with molecules possessing an extensive delocalised system of conjugated double bonds which results in a relatively rigid structure. This accounts for the intense fluorescence observed with molecules such as fluorescein, anthracene, perylene and other condensed ring aromatic structures. (It also accounts for the high value of the absorptivity in these compounds.)

If a molecule which absorbs uv radiation does not fluoresce it must have disposed of the excess energy by some other means in order to return to the ground state. In some cases the molecule may decompose because the energy it receives from a uv photon is comparable with the dissociation energy of chemical bonds. However there are other mechanisms by which the energy can be dissipated without the emission of radiation. Such processes are referred to as *radiationless transfer* of energy.

Radiationless transfer can occur in one of two ways. One way is by intra-molecular redistribution of the energy between the available electronic and vibrational states. This can be considered to take place in two stages, the first being called *internal conversion* and the second *vibrational relaxation* (the same process we met earlier involving the excited state following excitation).

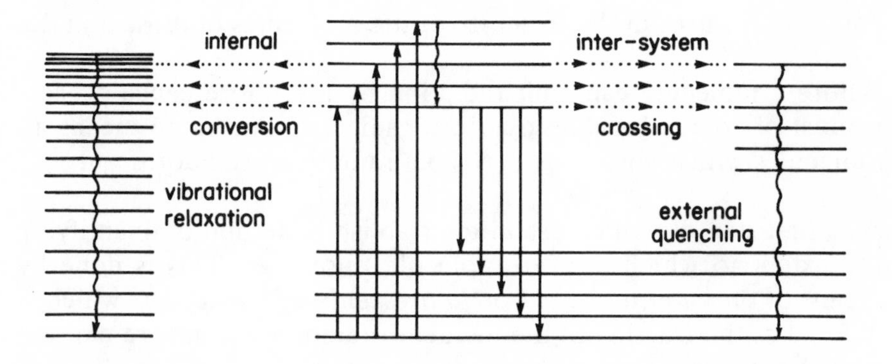

Fig. 1.4a. *Molecular excitation and routes for subsequent deactivation*

The other way is by a combination of intra- and inter-molecular energy redistribution. This too can be considered as a two stage process, with the first stage called *inter-system crossing* and the second *external quenching* (after vibrational relaxation has occurred).

Let's have a closer look at these processes with the help of the energy level diagram in Fig. 1.4a.

You will see that this diagram is an extension of that shown in Fig. 1.2a, and it shows the radiative transitions in the centre with the two possible radiationless transfer processes on either side. Notice that when we show the upper states of the vibration in the electronic ground state ($v'' > 5$), we have to take into account the fact that vibrational levels converge.

The internal conversion route on the left-hand side of the diagram shows the initial transfer from an electronic excited state to the electronic ground state. This places the molecule in a very high vibrational level of the electronic ground state from which it rapidly falls to the ground state by vibrational relaxation. The process is actually very complicated, but the details need not concern us. The important point is that the molecule finishes up in the ground state (electronic and vibrational).

The other route involving inter-system crossing is shown on the right-hand side of the diagram. The first stage sees the molecule crossing from the $v' = 0$ level of an electronic excited state, into a lower-lying excited electronic state. This is a special excited state which we will consider in more detail in Section 1.5. Suffice to say at this stage that, although the molecule can lose some energy by vibrational relaxation, it cannot now return to the ground state by emission of a photon because the spectroscopic transition is 'forbidden'. Instead, the molecule transfers its energy to other molecules by collision and other mechanisms. This is the process we have referred to as external quenching.

The conversion of electronic energy to vibrational energy and its subsequent degradation is much easier if the molecule is loose and floppy because it can reorient itself in ways which help to promote the internal transfer of energy. In this respect it behaves rather like a piece of soft rubber or plastic which can absorb the shock waves of a sudden impact such as a blow with a hammer, and distribute the energy throughout its entire bulk. A rigid block of metal or stone on the other hand simply transmits the shock waves to its surroundings. This behaviour is more akin to that of a rigid molecule which cannot make efficient use of internal conversion to return to the ground state and so is more likely to emit a photon. This somewhat crude picture may help you to appreciate the high fluorescence efficiency of many rigid planar molecules.

Not all rigid molecules are fluorescent, however, because the possibility of inter-system crossing is still open to them. This route to the ground state becomes particularly effective when the lowest electronic excited can be reached by an $n \rightarrow \pi^*$ transition from the ground state. (In the fluorescence process, however, it is more likely to be reached by a $\pi \rightarrow \pi^*$ transition to a higher energy excited followed by internal conversion.) The $n \rightarrow \pi^*$ excited state has a much longer lifetime than a $\pi \rightarrow \pi^*$ excited state and so there is a much greater chance of inter-system crossing taking place. Once this happens vibrational relaxation takes the molecule down to the $v' = 0$ level where it is trapped for long enough to be deactivated by collision or some other external quenching mechanism. This accounts for the very low fluorescence efficiency of compounds such as ketones and nitrogen heterocycles.

Inter-system crossing is also very efficient in compounds containing heavy atoms such as bromine and iodine for rather different reasons which we won't go into.

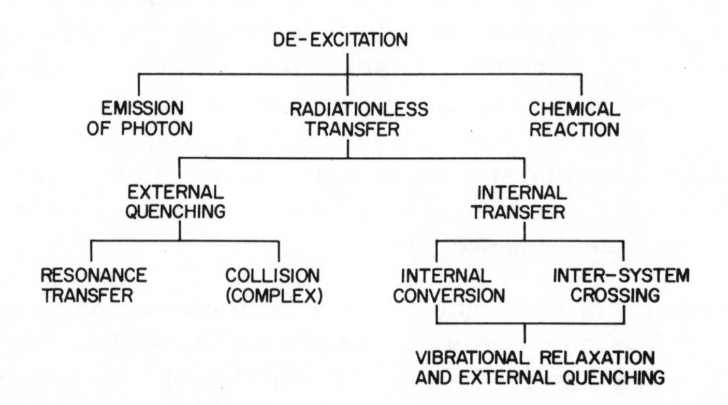

Fig. 1.4b. *Summary of the processes of molecular deactivation*

The relationship between the whole range of processes which lead to the de-activation of an electronically excited molecule are shown in Fig. 1.4b. This provides a 'hierarchy' of the terms used which you should find helpful for reference.

You will notice that quenching occurs on both sides of this diagram because it occurs with all types of excited molecules. Although it is most effective with those trapped in the 'special' states with no radiative pathway back to the ground state, it can still play a significant role in other cases. This has important implications in the use of fluorescence in analysis and we shall discuss it further in Part 3.

Let's have a look at some actual organic compounds and see if we can use the principles we have just discussed to predict whether or not they will fluoresce.

Π The molecular structures of six compounds are given below. Can you pick out three that are fluorescent?

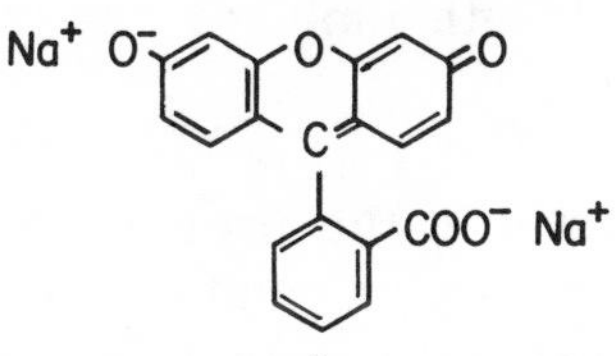

Na^+ O^- ... COO^- Na^+

A

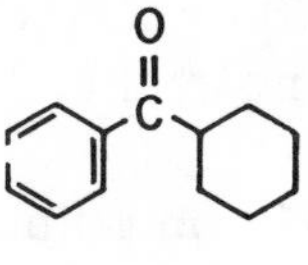

B

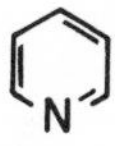

C

$CH_2CHCOOH$, NH_2

D

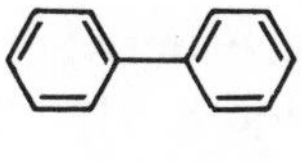

E

$CH_3COOC_2H_5$

F

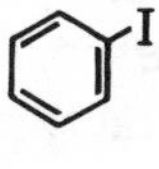

G

A is the fluorescein anion δ which is highly fluorescent due to the large conjugated planar system. You may have noticed that the ion does in fact contain a keto- group (the ring at the top right is quinonoid) but in this structure the $n \rightarrow \pi^*$ excited state is not the lowest excited state.

In B, cyclohexylphenyl ketone, the $n \rightarrow \pi^*$ excited state is lower than any $\pi \rightarrow \pi^*$ state and so it is not fluorescent.

Pyridine (C) also has a low-lying $n \rightarrow \pi^*$ state involving the lone pair electrons on the nitrogen and so is not fluorescent.

You might have thought that D was non-fluorescent for the same reason but in derivatives of pyrrole, the five-membered ring, the lone pair electrons are delocalised into the π-system of the ring and so no $n \rightarrow \pi^*$ transition is possible. Compound D is tryptophan, one of the few common amino-acids which is fluorescent. (Most of the others do not absorb above 210 nm.)

Biphenyl (E) is a large conjugated hydrocarbon with no complicating features and so it fluoresces. Although you might expect some internal rotation about the central bond, the molecule adopts a planar conformation to maximise conjugation across the two rings.

Ethyl acetate (F) does not fluoresce because it does not absorb uv radiation above 200 nm, the low wavelength limit for our instrumentation under normal circumstances.

Iodobenzene (G) is not fluorescent because the presence of the heavy iodine atom facilitates inter-system crossing and depletes the population of the initial excited state before it can emit a photon.

A, D and E are, therefore, fluorescent.

You may have gained the impression from this exercise that it is a difficult matter to predict whether or not a molecule will fluoresce from a simple examination of its structural formula. You would be quite correct! There is often a delicate balance between conflicting factors such as the presence of an extended π-system and features providing a mechanism for internal conversion or inter-system crossing and you can never be sure which will win. We might have correctly predicted that D is fluorescent while C is not if we had known about the delocalisation of the lone pair in the pyrrole ring. On the other hand we might have expected quinoline, to be non-fluorescent by analogy with pyridine. In fact it *is* fluorescent because the extension to the π-system gives rise to a wider range of π-orbitals and the $n \rightarrow \pi^*$ excited state is no longer the lowest excited state of the molecule.

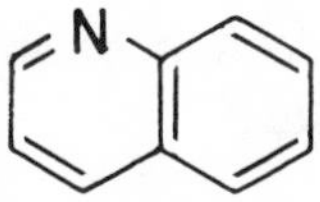

It is important to note that, in practice, the value of ϕ_f is affected by external factors, and in particular the solvent in which the fluorescent molecules are dissolved. The solvent therefore has a significant effect upon the intensity of fluorescence bands as well as their wavelength and general appearance.

SAQ 1.4a

The values of ϕ_f for three organic compounds A, B and C are 0.55, 0.91 and 0.12 respectively. Using this information only, which is the most strongly fluorescent?

SAQ 1.4b

State whether each of the following statements is true or false:

(*i*) Molecular fluorescence competes with two radiationless processes: Internal Conversion, Inter-system Crossing.

(*ii*) An excited molecule with a fused ring structure will readily lose its energy by internal conversion.

(*iii*) In order for inter-system crossing to be effective it must occur within 10^{-8} s of the molecule being excited.

(*iv*) A molecule in which inter-system crossing is very efficient is unlikely to be fluorescent.

(*v*) A molecule can pass from a $\pi \rightarrow \pi^*$ excited state to an $n \rightarrow \pi^*$ excited state by internal conversion.

(*vi*) Molecules in $\pi \rightarrow \pi^*$ excited states are more likely to undergo inter-system crossing than those in $n \rightarrow \pi^*$ states.

(*vii*) The fluorescence efficiency of a compound varies when it is dissolved in different solvents as a result of differences in the effectiveness of quenching.

(*viii*) Energy is removed very efficiently from the excited state of compounds containing bromine by inter-system crossing.

⟶

SAQ 1.4b (cont.)

(*ix*) De-excitation following inter-system crossing is achieved primarily by vibrational relaxation.

SAQ 1.4c

What three differences would you expect to find in the fluorescence spectrum of a compound showing vibrational fine structure if the solvent were changed from cyclohexane to ethanol?

SAQ 1.4c

1.5. THE PHOSPHORESCENCE MECHANISM

The process of inter-system crossing does not always lead to radiationless deactivation of the excited state. If we can prevent deactivation by collision it *is* possible to observe emission from this state. In order to achieve this we must reduce the rate at which collisions occur with the excited molecule so that the excited state is not quenched in spite of its long lifetime. This is most easily done by reducing the temperature which reduces the extent of molecular motion and thus the collision rate. Eventually, when the sample freezes, the molecules are locked in position in a rigid lattice and deactivation by collision ceases. In practice, the sample is cooled in liquid nitrogen to 77 K (−196 °C) at which temperature all common solvents are solid. The radiation emitted when the molecule returns to the ground state from the 'special' excited state is termed *phosphorescence* – another type of photoluminescence.

It is also possible to observe phosphorescence of some solids at room temperature, when the excited state molecules are not quenched by collision. However, the spectra of solid samples are rather more complicated to interpret for quantitative applications. Very recently, techniques have been devised to enable phosphorescence to be recorded from liquid samples at room temperature. This is likely to significantly enhance the use of phosphorescence in analysis in the near future.

We can show the transitions involved in phosphorescence on the same energy level diagram as the fluorescence if we add the vibrational levels of the second, lower-lying, excited state into which the molecule passes by inter-system crossing. This is done in Fig. 1.5a.

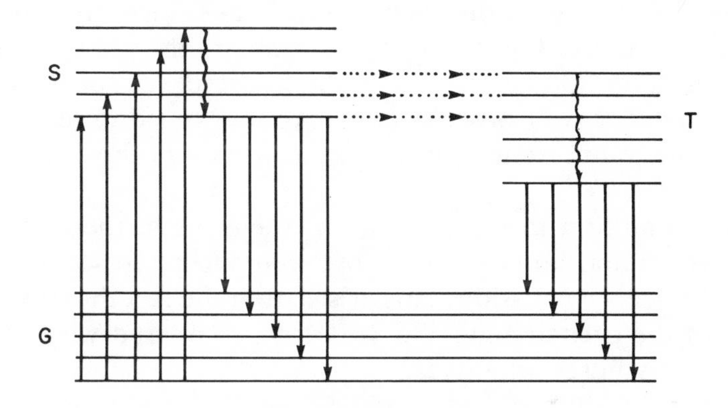

Fig. 1.5a. *Energy level diagram showing fluorescence and phosphorescence transitions*

The left-hand side of this diagram (states G and S) is identical with Fig. 1.2a and represents the fluorescence mechanism. The right-hand side of the diagram (states G and T) illustrates the phosphorescence emission, the excitation process being the same as for fluorescence. State T is the special excited state referred to in Section 1.4.

Π You should be able to deduce a very important property of phosphorescence from this diagram from your previous experience with fluorescence. What is it?

The phosphorescence emission is at longer wavelength than the fluorescence emission band. As before, this is apparent from the lengths of the arrows which show that the energy of the phosphorescence transitions is less than that of the fluorescence transitions.

Π What is implied about the fine structure (when observed) of the fluorescence and phosphorescence emission bands of a given molecule by Fig. 1.5a?

The spacing will be the same since the transitions terminate in the same series of vibrational levels in the ground state. There may, however, be differences in the relative intensities since different transitions are involved. Of course, if you compare the phosphorescence spectrum of the solid at 77 K with the fluorescence spectrum of the liquid, you may well observe more pronounced differences in the spectra due to differences in the physical state of the sample.

Π In addition to the increase in wavelength, there is another very important difference between phosphorescence and fluorescence emission. This arises from the fact that the lifetime of the excited state T is very long compared with that of S. Can you see what it is?

The phosphorescence is of much longer duration than fluorescence. Thus, when the exciting radiation is switched off the fluorescence intensity falls to zero very rapidly (within about 10^{-8} s). Phosphorescence emission however, continues long after the excitation radiation is removed, sometimes for a matter of seconds or even minutes. This enables us to observe phosphorescence without interference from fluorescence by a process known as *time resolution.* We shall discuss this further in Part 2.

There is also another type of long-lived photoluminescence which is observed under the same conditions as phosphorescence. This is called *delayed fluorescence* and arises as the result of a collision or some other process which imparts sufficient energy to the molecule in the excited state T, to raise it to a higher vibrational level from which it can cross back into the state S. It can then return to the ground state with the emission of a photon of the same wavelength as normal, 'prompt' fluorescence. The intensity of delayed fluorescence is very small compared with prompt fluorescence.

SAQ 1.5a

The energy level diagram given below shows some of the lower vibrational levels of a ground electronic state, G, and two excited electronic states, S and T.

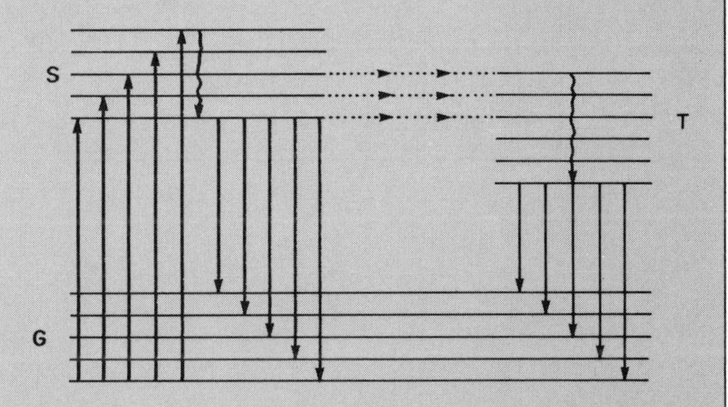

There are six different processes represented by arrows of various types. Label them A to F to correspond with the following descriptions:

A Excitation
B Fluorescence emission
C Phosphorescence emission
D Vibrational relaxation of the upper excited state
E Vibrational relaxation of the lower excited state
F Inter-system crossing

SAQ 1.5a

SAQ 1.5b

Place the following bands observed in the photoluminescence spectrum of a single compound in order of increasing wavelength:

fluorescence emission
fluorescence excitation
phosphorescence emission

SAQ 1.5c

The diagram below shows how the intensity of photoluminescence emission, excited by a single 'pulse' of uv radiation, varies with time. Of the two curves (full line and dotted) which corresponds to fluorescence and which to phosphorescence?

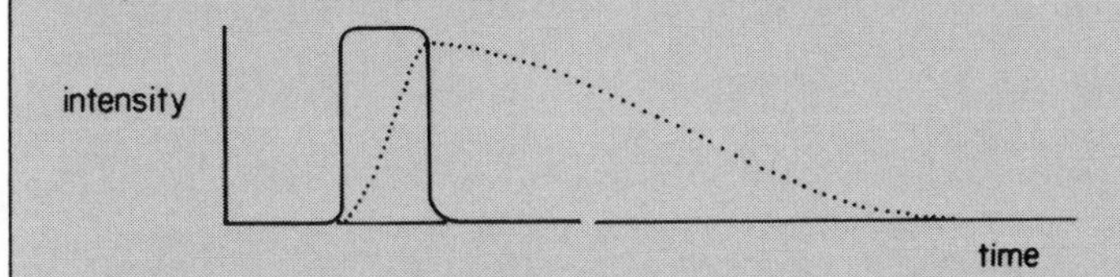

Indicate on the diagram the time at which

(*i*) the pulse was switched on, by t_1;
(*ii*) the pulse was switched off, by t_2;
(*iii*) the phosphorescence could be measured without interference from fluorescence, by t_3.

SAQ 1.5d

What radiation processes compete with

(*i*) fluorescence emission
(*ii*) phosphorescence emission

as the means of deactivating an excited molecule?

SAQ 1.5e

How could you check whether long-lived emission from a sample frozen at 77 K was phosphorescence or delayed fluorescence?

SAQ 1.5e

1.6. SINGLET AND TRIPLET STATES

In order to explain why it is that the special excited state associated with phosphorescence has such a long lifetime we need to enquire a little more closely into the nature of electronic states in general. The vast majority of organic molecules contain an even number of electrons and, in the ground state, these are normally grouped in pairs, each pair occupying a single molecular orbital. According to the Pauli Exclusion Principle, two electrons can occupy the same orbital only if their spins are 'paired', that is, when the magnetic fields associated with the spinning electrons are in opposite directions and therefore cancel out. (You may be more familiar with the Pauli Principle in the context of atomic structure and the building up of the Periodic Table.)

Since the spin of every electron in the molecule is cancelled out by the spin of its companion in the same orbital, the total electron spin of the ground state molecule is zero and the state is said to be a 'singlet state'. When the molecule is excited, one of a pair of electrons moves into an orbital of higher energy. Its spin may still be paired with that of the electron left behind and so a whole series of excited states exists in which the total electron spin is still zero. Like the ground state, these excited states are also singlet states. The ground state is often given the symbol S_0, the first excited state S_1, and higher singlet states S_2, S_3 etc.

The differences between these electronic states become clearer if we draw diagrams to represent their electronic configurations (ie the way in which the electrons are distributed amongst the available orbitals). The diagram below shows the electronic configuration of a simple molecule like propanone which has a π-system due to the presence of a C=O double bond, and lone pair electrons on the oxygen atom.

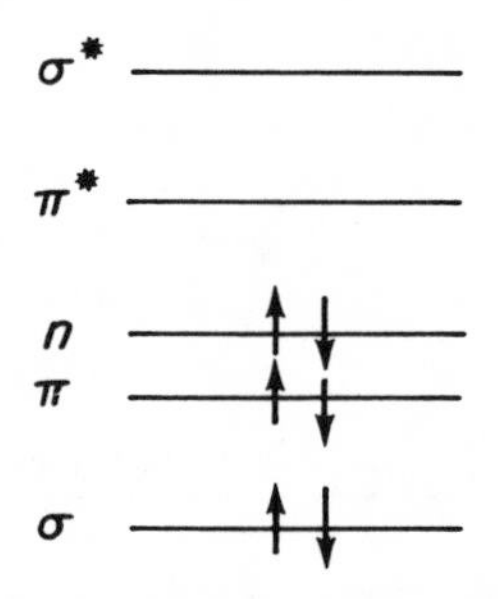

The horizontal lines represent the energies of the three highest occupied and the two lowest unoccupied orbitals. The arrows represent electrons with their spins paired in each occupied orbital which gives the ground state configuration. The occupied orbitals are designated σ (bonding), π (bonding) and n (non-bonding) in ascending order of energy. The unoccupied orbitals are then π^* and σ^* (both anti-bonding).

Π See if you can construct similar diagrams to represent the electronic configurations of the same molecule in the two excited singlet states, S_1 and S_2. Remember that the $n \rightarrow \pi^*$ transition is of lower energy than the $\pi \rightarrow \pi^*$ for this molecule.

You should have produced something similar to the diagrams shown here.

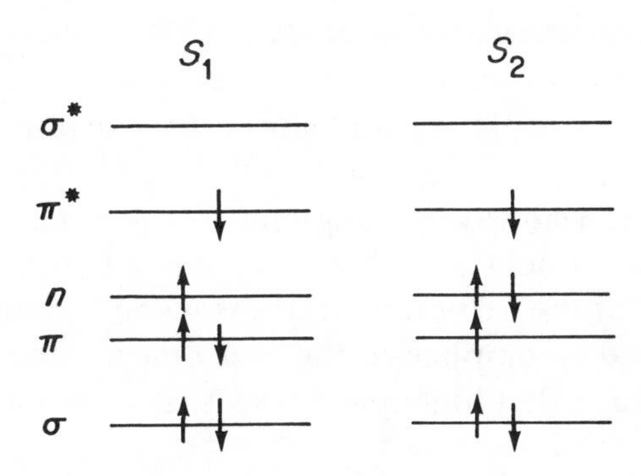

[S_1 is the lowest $n \rightarrow \pi^*$ excited state.]

[S_2 is the lowest $\pi \rightarrow \pi^*$ excited state.]

In the excited state, the two electrons occupying separate orbitals are no longer restricted by the Pauli principle and so it is quite possible for their spins to be 'parallel' (ie aligned in the same direction). The series of excited states in which two electron spins are not paired have an overall spin momentum and therefore exhibit magnetic properties (they are 'paramagnetic'). In a magnetic field, each state is found to split up into three sub-states of slightly different energies. These states are therefore referred to as 'triplet states' because they have a 'multiplicity' of three. We can now see why the (diamagnetic) singlet states are so called because they remain single in an applied magnetic field. The lowest triplet state, which is of course an excited state of the molecule, is often labelled T_1 and higher triplet states T_2, T_3, etc. Note that for every singlet excited state there is a corresponding triplet state differing from it only in the direction of the spin of one electron. This is sufficient to affect the energy slightly and the triplet state is always of lower energy than the corresponding singlet state. (This is a consequence of Hund's 'rule of maximum multiplicity' which again you may have met in the context of atomic structure.)

Π Turn back to Fig. 1.5a if you cannot recall the energy level diagram showing electronic states labelled G, S and T. We have introduced new symbols in the preceding two paragraphs, which ones correspond to G, S and T?

$$G = S_0 \qquad S = S_1 \qquad T = T_1$$

(the original choice of letters was not entirely arbitrary!)

Let's see if you can now draw diagrams to show the electronic configurations of the triplet states T_1 and T_2 which correspond to S_1 and S_2. The outline of the orbital energies is given to help you. The σ^* orbital is retained to emphasise the fact that we are looking at the lowest excited state, that higher states exist and what these might be.

	S_1	T_1	S_2	T_2
σ^*				
π^*				
n				
π				
σ				

Your final array of excited state configurations should look something like this:

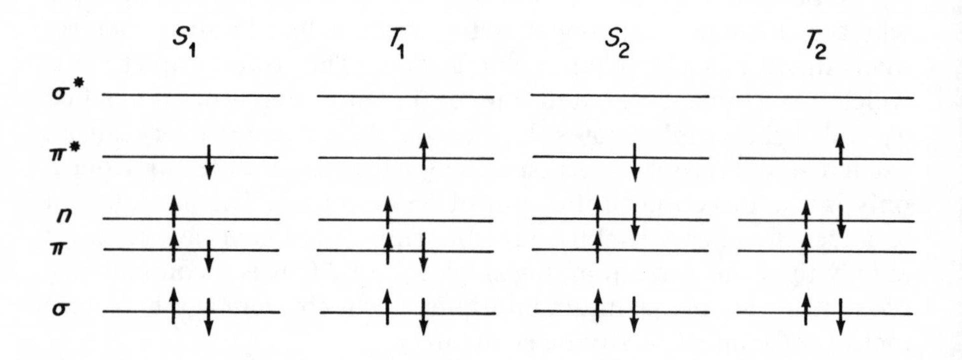

I hope you got at least the singlet states correct because we have already done them once.

These diagrams are very approximate representations of the actual orbital energies and so it is not always possible to see whether one state is of higher energy than another. It is, in fact, very difficult to obtain the information needed to draw accurate diagrams because the energy of one orbital depends on whether it contains one or two electrons and also on which other orbitals are occupied. However, our simplified diagrams will give you the general idea and you don't need to worry about the finer points. The energy of a molecule in an excited state is notionally obtained by adding up the energies of all the electrons (not orbitals since empty orbitals don't have any energy). In practice the energies of excited states are determined from spectra which give the energy gap between ground and excited states. To make life easy, the energy of the ground state is always taken as zero.

Excitation of ground state molecules involves transitions such as $S_0 \rightarrow S_1$, $S_0 \rightarrow S_2$, $S_0 \rightarrow S_3$ etc. which can be achieved by the absorption of uv radiation of progressively shorter wavelength. Transitions from the ground state to triplet excited states, however, cannot occur because they violate one of the laws of quantum mechanics. This is a 'selection rule' which states that the electron spin can not change during a transition associated with the absorption or emission of radiation. Consequently, molecules can arrive in triplet states only by inter-system crossing from a singlet state or, in some cases, by some other non-radiative process such as a chemical reaction. Similarly, since singlet – triplet transitions are 'forbidden' in both directions, a molecule arriving in the lowest triplet state, T_1, cannot return to the ground state, S_0, by emitting a photon.

In practice, the selection rule is not quite rigorous and there is a small probability that such a forbidden transition can take place. Thus, a molecule crossing from S_1 to T_1 can *eventually* return to the ground state with the emission of radiation provided of course that it is not deactivated by some other process first. It is this principle that accounts for the long lifetime of the triplet state and the associated phosphorescence emission.

A similar argument accounts for the longer lifetime of $n \rightarrow \pi^*$ excited states compared with $\pi \rightarrow \pi^*$ states. In this case the selection rule is even less rigorously obeyed. The low absorptivity of about $10\ \mathrm{l\ mol^{-1}\ cm^{-1}}$ for the $n \rightarrow \pi^*$ band compared with the value of about 10 000 l mol^{-1} cm^{-1} for the $\pi \rightarrow \pi^*$ band is a measure of the greater probability of the $\pi \rightarrow \pi^*$ transition. (The corresponding value for a singlet-triplet transition is thought to be of the order of 0.01 l mol^{-1} cm^{-1}.)

Now that you have got the idea of singlet and triplet states we can define the terms we introduced in Section 1.4 in an alternative, more rigorous, way:

Internal Conversion – a radiationless transfer between electronic states of the same multiplicity (eg singlet↔singlet

Inter-system Crossing – a radiationless transfer between electronic states of different multiplicity (eg singlet↔triplet)

SAQ 1.6a

Which of the statements (*i*) to (*v*) correctly completes the following sentence?

Transitions from T_1 to S_0 are forbidden because

(*i*) the state is depopulated by inter-system crossing before the transition can occur.
(*ii*) T_1 is of lower energy than S_1.
(*iii*) they involve a change in the electron spin of the molecule.
(*iv*) the spin of the T_1 state is zero.
(*v*) the probability of singlet-triplet transitions is very low.

Summary

Fluorescence and phosphorescence are forms of luminescence in which light or uv radiation is emitted by a compound following excitation with radiation of shorter wavelength in accordance with Stokes' Law. Phosphorescence has a longer wavelength than fluorescence and also has a longer lifetime. Both processes involve the excitation of molecules into electronic excited states and the emission of luminescence competes with other processes by which the molecule can lose its excess energy. The energy can be transferred directly to other molecules by a process known as quenching or dissipated through the vibrational motion of the molecule. Quenching depends on collisions between molecules and the vibrational process is facilitated by internal transfer of energy by internal conversion and inter-system crossing as a result of which the molecule passes over into a lower-lying electronic state. In the case of phosphorescence, it is necessary to prevent quenching by observing the sample in the solid state or protected from collisions in some other way.

There are two different types of fluorescence spectra, the excitation spectrum (which is identical with the absorption spectrum) and the emission spectrum. These bear a mirror-image relationship to each other which is particularly apparent when the bands show vibrational fine structure. The separation between the band maxima is known as the Stokes' shift. The phosphorescence spectrum has a similar vibrational fine structure to the fluorescence spectrum though in practice this is often obscured because the sample is in a different physical state. The transitions involved in these spectra can be conveniently shown on a Jablonski diagram which also identifies the specific vibrational levels involved by the use of the vibrational quantum number as a label. The 0,0 transition is common to both spectra but usually appears at longer wavelength in the emission spectrum because of solvent interaction.

The fluorescence efficiency of a compound is defined as the fraction of the incident radiation which is re-emitted as fluorescence and given the symbol ϕ_f. A high value of near 1.0 is generally observed with molecules having large, planar conjugated systems which are relatively rigid. More flexible molecules are more likely to have low values of ϕ_f while molecules whose lowest excited state is achieved by an $n \rightarrow \pi^*$ transition or which contain heavy atoms like bromine and iodine are usually non-fluorescent.

The excited state associated with phosphorescence is a triplet state which has a very long lifetime (typically 10^{-3} s) compared with the singlet state (10^{-8} s) because the transition to the (singlet) ground state is spin-forbidden. This accounts for the long lifetime of phosphorescence emission. The triplet state is also involved in another type of long-lived luminescence called delayed fluorescence.

Objectives

You should now be able to:

- describe the general appearance of fluorescence and phosphorescence spectra;
- distinguish between fluorescence and phosphorescence;
- describe the molecular processes associated with fluorescence and phosphorescence emission;
- account for the long lifetime of phosphorescence;
- recognise structural features in a molecule which are likely to enable fluorescence or phosphorescence to occur;
- compare the fluorescence efficiencies of different molecules in terms of their structures;
- describe processes which may in practice reduce the fluorescence efficiency of a given molecule.

With this general background you are now in a position to study the application of fluorescence and phosphorescence to problems in analytical chemistry. Before doing so, however, we shall first have a look at the instrumentation required for the observation of photoluminescence. This is the subject of Part 2 of this Unit.

2. Instrumentation

2.1. INTRODUCTION

The instrumental requirements for measuring fluorescence are:

(*i*) a source of uv radiation;

(*ii*) a wavelength selector to enable the operator to choose the wavelength of the radiation to be used to excite the sample molecules;

(*iii*) an optical system to direct the exciting radiation onto the sample and to collect the emitted radiation;

(*iv*) a second wavelength selector to enable the operator to observe the radiation emitted at a particular wavelength;

(*v*) a sensitive detector to respond to the emitted radiation;

(*vi*) a read-out system to record the intensity of the fluorescence emission.

Much of this instrumentation is very similar to that used in uv absorption spectroscopy which you will probably already have met. In this Unit we need only to highlight the features of those components common to both techniques which are of particular importance in fluorescence work and to describe in more detail the components which are specific to fluorescence. In practice many commercial uv spectrometers are provided with accessories which enable them to be used to measure fluorescence. Whilst not as efficient as purpose-built instruments, these accessories provide an economic alternative with adequate performance for many applications of fluorescence.

Π Can you remember the general layout of a conventional uv spectrometer? See if you can draw a block diagram of it.

Your diagram should look something like Fig. 2.1a:

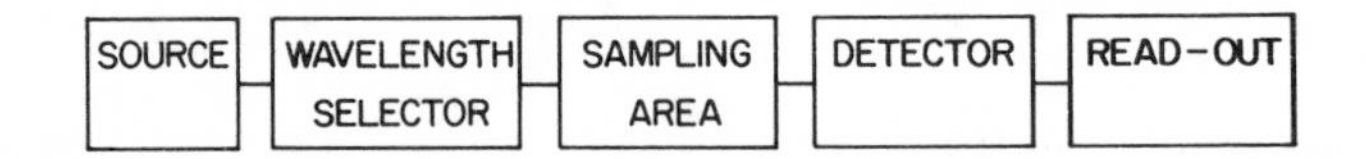

Fig. 2.1a. *Layout of a typical uv absorption spectrometer*

Possibly you did not show the sampling area as a separate block and you may have named the principal components used in other blocks rather than the function of the block.

Π What are these components when the instrument is set up to measure absorbance at 300 nm?

Source: deuterium lamp. You may have called this a hydrogen lamp; although this is possible, in modern instruments deuterium is now universally used in place of hydrogen. You might also have chosen the tungsten/halogen/quartz lamp which would also be successful at this wavelength. However, below 300 nm the deuterium lamp becomes progressively more satisfactory. The ordinary tungsten filament lamp is *not* suitable because its emission does not extend significantly below the visible range and its glass envelope cuts out radiation below 320 nm.

Wavelength selector: grating monochromator (prisms are now virtually obsolete) or filter (the interference type is preferable).

Detector: photo-cell or photomultiplier.

Read-out: millivoltmeter (analogue or digital) or printer or chart recorder.

You are probably aware that there is a variety of instruments available in the uv and visible regions of the spectrum using different combinations of these components. The same is true of fluorescence instruments.

Π Let's now consider what changes need to be made to this basic spectrometer layout in order to produce an instrument capable of measuring fluorescence. There are three chief alterations, one of which should be obvious from the list of instrumental requirements given at the start of this section. What is it?

Two wavelength selectors are required, one on each side of the sample. The other two changes concern the source and the sample geometry and are less apparent so let's look at these first.

2.2. RADIATION SOURCES

One of the most important differences between absorption and fluorescence instruments concerns the source. In absorption work the instrument is used to measure the transmittance or absorbance of the sample. This involves taking the ratio between the intensity of the radiation passed by the sample and by a blank solution when placed in turn in the beam of radiation. The absolute magnitude of the intensities is not important provided they are large enough to produce a signal which is well above the noise level of the detector system. In fact, an intense source is not desirable because of the possibility of molecular photo-decomposition and the problems caused by fluorescence in an absorption instrument. A typical deuterium lamp, therefore, dissipates less than 10 watts.

By contrast, the intensity of fluorescence is directly proportional to the intensity of the incident radiation so that there is a much to be gained from using a source of high power – provided, of course, that the sample is not photodecomposed. It is also important that the output of the source should be constant so that the fluorescence readings do not vary with time. A 'continuous' source (ie one emitting radiation over a wide range of wavelength) is generally preferred so that the operator has a free choice of wavelength. The type of deuterium lamp used in uv spectrophotometers cannot easily be adapted to give high power output so an alternative has to be found.

2.2.1. The Xenon Arc

At present, the most commonly used source for fluorescence measurements is the high-pressure xenon arc which is usually operated at 150 watts. Although xenon is a monatomic gas and hence would normally be expected to emit a line spectrum, under the conditions prevailing in the arc (which operates at several amps) the atomic lines are broadened to such an extent that they overlap. For this and other more complex reasons the xenon arc is essentially a continuous source. A typical output curve for a lamp of this type is shown in Fig. 2.2a.

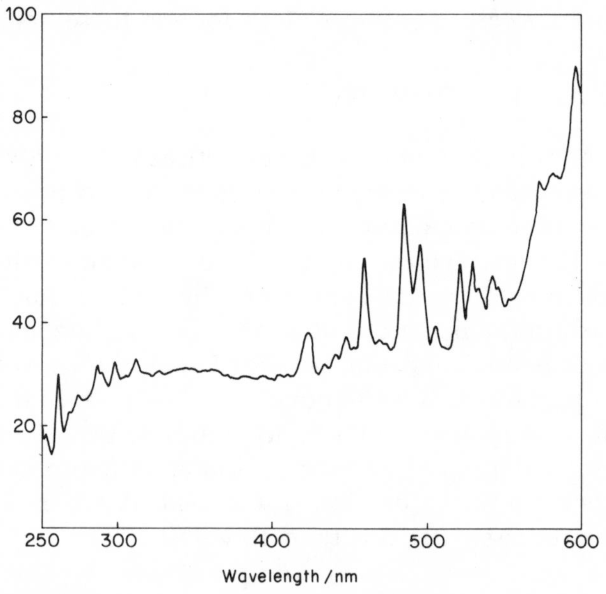

Fig. 2.2a. *Spectral emission typical of a Xenon Arc (courtesy of Perkin–Elmer)*

Considerable care has to be exercised in the handling and use of high-pressure xenon lamps.

Π From the properties of this source outlined above, try to identify two or possibly three potential hazards to which the user might be exposed.

1. If the lamp is dropped it is liable to explode scattering fragments of silica around the laboratory. This arises from the very high pressure inside the envelope (amounting to several atmospheres). This hazard is present only when a lamp is being changed or handled outside the instrument and so does not occur during normal operation.

2. The lamp emits very intense uv radiation which is damaging to the eyes. You should never look at the unshielded lamp when it is on. In practice, under normal operating conditions it is not possible to do this and the hazard would only arise if you had to remove a cover in order to adjust the position of the lamp. You should also take care not to look directly into the incident beam when adjusting accessories. This could be dangerous if uv radiation were present.

3. A less familiar hazard which you may not have encountered before is that very intense sources of short wavelength uv radiation convert some of the oxygen in the atmosphere into ozone. This is extremely toxic and great care must be taken not to expose the operator to it even at very low concentrations. This hazard was formerly avoided by careful attention to ventilating the source unit to the outside atmosphere, but with modern instruments a de-ozoniser system is built into the lamp housing which takes care of the problem at source.

   (A similar problem arises with some photocopying machines.)

Although it does not actually constitute a hazard, if you ever have to handle a uv lamp (of any type) you should be careful not to leave fingerprints on it. If you do, when you switch the lamp on, the fingerprints will be burned into the silica envelope and significantly reduce its output. The usual remedy is to wash off the fingerprints or other greasy marks with ethanol.

2.2.2. Mercury Vapour Lamps

The only other source you are likely to come across in connection with fluorescence work is the mercury vapour lamp. Experimentally it is easier to produce high intensity uv radiation from discharge lamps containing the vapour of metals such as mercury. These lamps produce line spectra, of course, but for many purposes this is not a serious disadvantage. The most intense line of mercury is at 254 nm, well down in the uv, but its intensity is often seriously reduced by impurities in the quartz envelope. In high pressure versions of the lamp, which can be operated at higher power, it is virtually absent due to self-absorption. However, there are other useful lines at 313, 365, 405, 436 and 546 nm. Any of these can be isolated by a suitable interference filter to give essentially monochromatic exciting radiation. It is perhaps worth mentioning that the familiar 'fluorescent tube' used for general lighting purposes is in fact a mercury discharge lamp in a tube coated with a fluorescent powder which emits white light when excited by uv radiation from the mercury atoms.

SAQ 2.2a

Explain why the high pressure xenon arc is more satisfactory as a source for fluorescence work than either the tungsten filament lamp, the deuterium lamp or the mercury discharge lamp. When might the mercury lamp be a convenient alternative?

SAQ 2.2a

2.3. SAMPLE GEOMETRY

Another major difference between fluorescence and absorption instruments lies in the geometry of the sampling area. For absorbance measurements the beam travels straight across the sample position and we are concerned only with the reduction in its intensity after it has passed through the sample. In fluorescence measurements we have to observe the fluorescence emission in the presence of the much stronger exciting radiation. Although this is possible with the 180° 'straight through' geometry because the incident and emitted radiation is of different wavelength, the unabsorbed incident radiation passes into the monochromator and gives rise to serious stray light problems.

A much more satisfactory arrangement is to view the fluorescence at 90° to the incident beam. This is possible because fluorescence is emitted in all directions from the sample. Direct light cannot then enter the monochromator though some radiation at the incident wavelength always accompanies the fluorescence radiation as a result of scattering. In a perfectly clear solution this is due to Rayleigh scattering from solvent molecules which is often of comparable intensity with the fluorescence of the solute. If the fluorescent species is at very low concentration, Rayleigh scattering becomes more troublesome and a further problem arises from the Raman scattering of the solvent. Although this is very much weaker than Rayleigh scattering, it occurs at longer wavelength and in many cases overlaps and interferes with the emission band of the fluorescent species. If the sample contains suspended particles these make a further contribution to the scattering at the exciting wavelength through Tyndall scattering. This can be far more intense than Rayleigh scattering if the sample is at all turbid and so can give rise to high stray light levels. These scattering processes and their significance in analysis are discussed more fully in Part 3.

The use of frontal illumination, in which the emitted radiation is observed through the same face of the cell through which the exciting radiation is incident, can reduce the problem of scattering from turbid and solid samples. It is also useful for strongly absorbing samples in which the exciting radiation does not penetrate more than a millimetre or so into the solution. These three sampling geometries

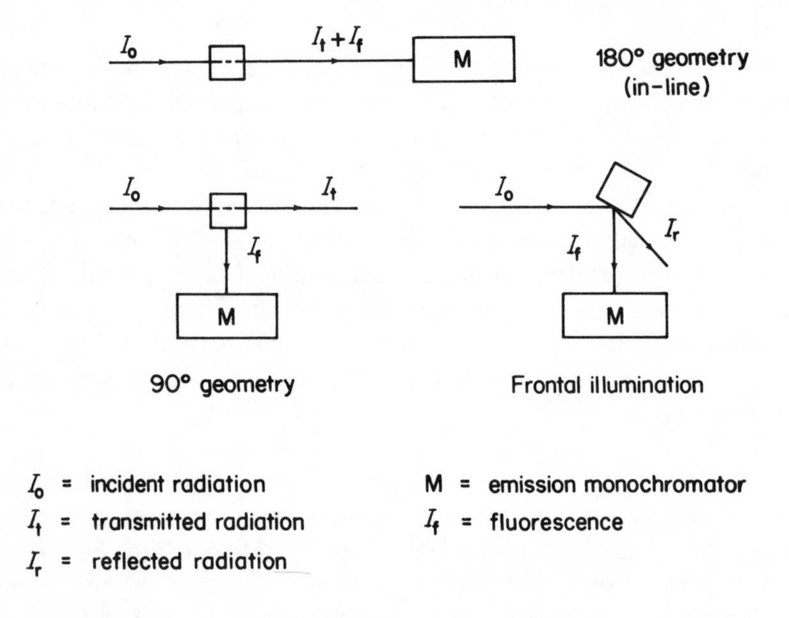

Fig. 2.3a. *Three alternative sampling geometries*

are shown in Fig. 2.3a. Most fluorescence instruments adopt the 90° geometry though the other configurations are often available as alternatives.

Only the fluorescence which enters the monochromator at 90° is shown in Fig. 2.3a. This is only a small part of the total emission which occurs in all directions. Instruments can in fact be made more sensitive by the use of a more efficient optical system to collect a greater proportion of the emission. A simple arrangement is shown in Fig. 2.3b where a concave mirror is placed at 90° to the incident beam but opposite to the monochromator to focus reflected radiation on to the slit of the monochromator. An alternative arrangement used by some manufacturers is shown in Fig. 2.3c. Here a plane mirror is placed opposite to the incident beam to give it a second pass through the sample.

Π Which of these arrangements would you imagine to be the more effective? What increase in the signal might be expected?

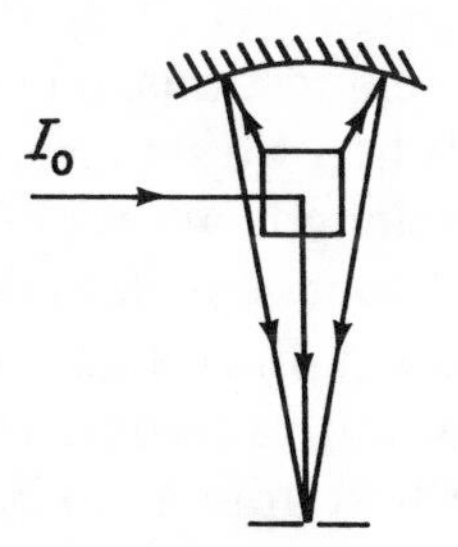

Fig. 2.3b

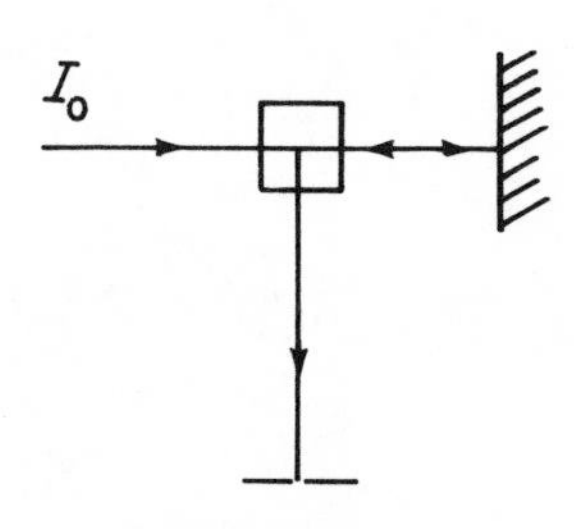

Fig. 2.3c

There really isn't very much in it but the arrangement in Fig. 2.3c introduces fewer complications. Both systems would increase the signal by a factor of about 2.

The arrangement in Fig. 2.3b suffers from the disadvantage that much of the reflected radiation would pass through the sample for a second time and would be subject to selective absorption and scattering. There would also be a chance of picking up reflected radiation from the cell walls and other components which would add to the stray light. A further contribution to the stray light and wavelength dependent intensity variations would arise from any accumulation of dust on the mirror. This deterioration in optical surfaces is inevitable in a normal laboratory atmosphere and its effect is greatly enhanced as the number of optical components is increased.

The 90° configuration is of great importance in spectrofluorimetry and makes a very significant contribution to the high sensitivity of the technique. This will be discussed more fully in Part 3 but for the time being we will demonstrate the advantage of 90° geometry quantitatively in the following SAQs.

Before attempting to answer the questions, let us make sure we know what is meant by stray light. When we set a monochromator to pass radiation of a particular wavelength we wish to prevent radiation at a range of higher and of lower wavelengths from reaching the detector. If the instrument has a stray light level of 0.1% we can only stop 99.9% of the unwanted radiation reaching the detector. This isn't as good as might at first appear to be the case.

SAQ 2.3a

A spectrometer has a stray light level of 0.01%. In a particular experiment the value of I_0 was 1000 arbitrary units. The sample absorbed 1% of the incident radiation and scattered a further 1%. Calculate the value of I_f which is equal to the stray light level, using 180° geometry.

SAQ 2.3b

Calculate the value of I_f which is equal to the stray light level, using 90° geometry for the same instrument as in SAQ 2.3a. Assume that 1% of the total scattered light enters the monochromator.

Hence estimate the gain in sensitivity of the 90° configuration over the 180° or this sample.

2.4. CELLS FOR FLUORESCENCE

Another advantage of 90° sample geometry is that the sample can be examined in cells of the same shape and size as those used in absorption work.

Π There is however a fundamental difference between absorption and fluorescence cells. Perhaps you can see what this might be – think about it for a moment.

The standard absorption cell is usually 1 cm square and 4–5 cm high. It is made of glass or silica with opposite faces polished, the other faces having a 'frosted' finish. For fluorescence work with 90° geometry it is obviously necessary to have two *adjacent* sides polished. In practice fluorescence cells usually have all four faces polished and so for very precise quantitative work it is important to ensure that they are always inserted into the cell holder the same way round. As with absorption cells, they are usually made of glass or silica using a fused method of fabrication. For some purposes plastic (polystyrene) cells may be used which are cheaper than glass or silica but lack the dimensional stability of the traditional cells.

The requirements on the quality of materials used for fluorescence are much more stringent than for absorption. In particular they must have as low a natural fluorescence as possible. For example, synthetic fused silica is always used instead of natural quartz because it is free from fluorescent impurities. It is also necessary for the operator to exercise great care in handling fluorescence cells to avoid scratching the optical surfaces or contaminating them with fingerprints. Cells must always be washed out carefully after use and replaced in properly designed storage cases.

SAQ 2.4a

Suggest a situation in which it would *not* be possible to use

(*i*) a glass cell
(*ii*) a silica cell
(*iii*) a polystyrene cell

for making fluorescence measurements on a liquid sample.

SAQ 2.4b

How would

(*i*) a scratch, and
(*ii*) a fingerprint

on the optical surfaces of a cell affect the measurement of fluorescence?

SAQ 2.4b

2.5. TYPES OF FLUORESCENCE INSTRUMENT

Π We have now identified the three major differences between fluorescence and absorption instruments. What are they?

A more powerful source is used, and the emitted radiation is viewed at 90° to the incident radiation. Two wavelength selectors are required to control the wavelength of the incident and emitted radiation independently.

Π You should now be able to draw the block diagram of a typical fluorescence spectrophotometer. Have a go!

Check your diagram against the one shown in Fig. 2.5a. Well done if it is recognisably the same – you are obviously getting the hang of this instrumentation business! If not, don't despair; you should be able to account for the differences quite easily.

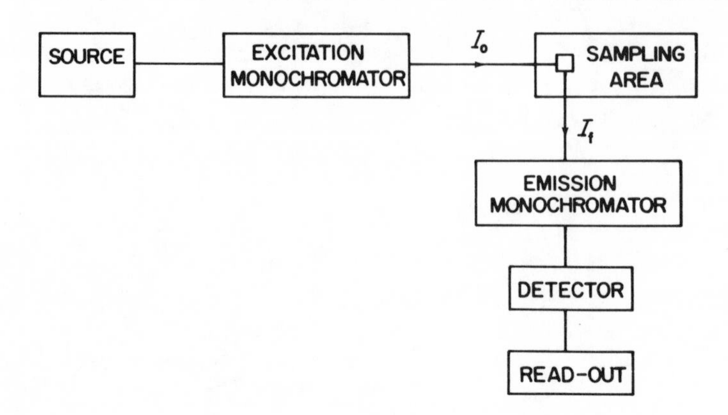

Fig. 2.5a. *Layout of a typical fluorescence spectrometer*

We are now in a position to have a look at the various types of fluorescence instrument which are likely to be found in analytical laboratories.

2.5.1. Dual Monochromator Spectrofluorimeters

The most versatile and therefore the most expensive fluorescence instrument is the dual monochromator spectrofluorimeter. (The terms 'fluorescence spectrometer' and 'spectrofluorimeter' are equivalent – the American spelling of the latter is 'spectrofluorometer'). This has two grating (formerly prism) monochromators and gives the operator a completely free choice of wavelength for both excitation and emission. It also allows the operator to record both the excitation and emission spectrum of a sample.

The performance of grating monochromators is determined by a number of factors which are listed in the manufacturer's specification and include the following:

(*a*) light-gathering power – usually quoted as the 'aperture' or 'f-number' of the optical system
(*b*) resolving power
(*c*) dispersion
(*d*) stray light

Π Which of these factors would you expect to be of *greater* importance in fluorescence than in absorption spectroscopy?

(*a*) and (*d*).

We noted earlier that the use of a high power source is beneficial in fluorescence work since it results in a higher fluorescence intensity which makes it easier to detect low concentrations of the fluorescent species. A further enhancement of the sensitivity is achieved by optimising the collection of fluorescence emission from the sample. The use of a high aperture monochromator provides a further improvement in optical efficiency and therefore in sensitivity. This method is in fact preferable to the use of mirror systems to optimise light gathering from the sample which, as we have seen can cause other problems. A typical fluorescence spectrophotometer has an aperture of $f/2$ compared with $f/4$ for a comparable absorption instrument. (As with camera lenses, the lower the f-number the higher the aperture.)

Since the detail in fluorescence spectra is similar to that in absorption spectra (the excitation spectrum being identical) a similar specification in terms of both dispersion and resolving power is appropriate. The stray light level however is much more important in fluorescence work where we are often trying to measure low signals from the analyte against a relatively high background signal.

Let's now take a look at the way in which this instrument is used to obtain the emission spectrum of a compound. For this purpose we have to scan the emission monochromator whilst the incident radiation is at a fixed wavelength. We shall get the highest emission intensity if we excite the sample with radiation at the wavelength of maximum absorption which we shall probably know in advance. Take, for example, the compound phenanthrene (an isomer of anthracene) which has its maximum absorbance at 354 nm. We therefore set the excitation monochromator to this wavelength and scan the emission monochromator over the entire range of wavelength over which emission is observed. On most instruments this also sets the chart recorder running and we obtain a trace of emission intensity versus wavelength. The spectrum is usually presented with the wavelength increasing from left to right – the same convention as that normally adopted for absorption spectra).

The wavelength scale is often added as a series of 'blips' at 10 or 20 nm intervals with a second marker pen. The operator is required to enter the wavelength against at least one of these marks at the time the spectrum is run so that it is properly calibrated. In modern instruments, however, increasing use is being made of microprocessors to control the instrument and to display results. Instead of a pen recorder, these instruments usually output the results to a printer/plotter which not only draws the spectrum but also encloses it in axes and adds the wavelength scale automatically.

When the sample is a clear solution it is convenient to scan from just below the excitation wavelength up to the point at which the emission intensity falls to the baseline level. This records the Rayleigh scattering as well as the fluorescence and avoids the possibility of missing any features close to the Rayleigh peak. This is not advisable however if the sample is scattering strongly or if the instrument is being operated at high gain to observe very weak fluorescence. In this case the scan should be started as close to the Rayleigh peak as possible without sending the recorder pen too far off scale.

The emission spectrum of phenanthrene recorded in this way with a printer/plotter on a modern instrument is shown in Fig. 2.5b.

Π Where is the Rayleigh peak in this spectrum?

At 354 nm, the wavelength of excitation; at this point in the spectrum both monochromators were set to the same wavelength.

Π By analogy with the preceding discussion, describe how you would obtain the excitation spectrum of phenanthrene shown in Fig. 2.5c. The same considerations apply to the Rayleigh peak.

Set the emission monochromator to 400 nm (the wavelength of maximum emission) and scan the excitation monochromator from the low wavelength limit (usually near 200 nm) up to, or just beyond, the emission wavelength depending on the amount of scatter or the

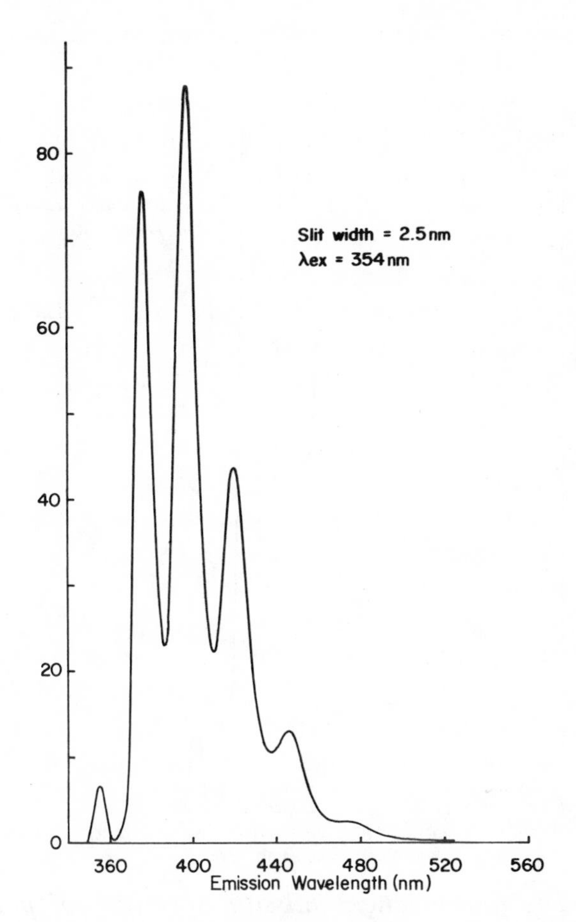

Fig. 2.5b. *The fluorescence excitation spectrum of phenanthrene (100 mg dm^{-3})*

relative intensity of fluorescence to Rayleigh scattering. Remember to calibrate the wavelength scale before removing the spectrum from the recorder if the instrument does not do this automatically. (Our instrument does.)

Π Where is the Rayleigh peak in Fig. 2.5c?

At 400 nm. Notice that the Rayleigh peak comes at the long wavelength end of the excitation spectrum and the low wavelength end of the emission spectrum.

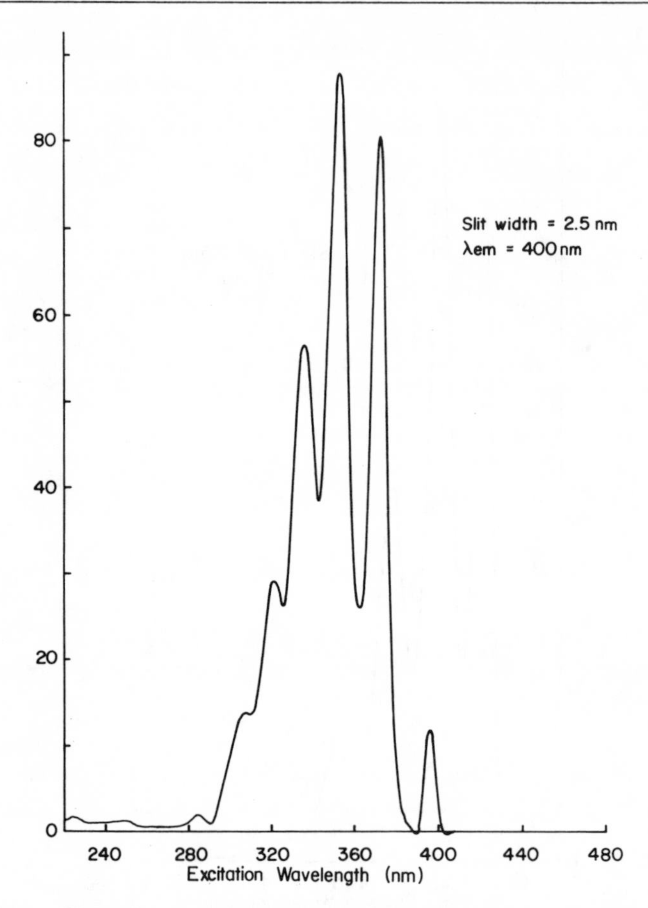

Fig. 2.5c. *The fluorescence emission spectrum of phenanthrene (100 mg dm^{-3})*

2.5.2. Filter/Monochromator Instruments

A considerable saving in the cost of a spectrofluorimeter can be made if one of the monochromators is replaced with a band-pass filter. Modern interference filters have a high peak transmission (typically 60%) and a band-width comparable with that of a small grating monochromator (5 nm). The transmission curve of such a filter is shown in Fig. 2.5d.

Interference filters have a considerable advantage over grating

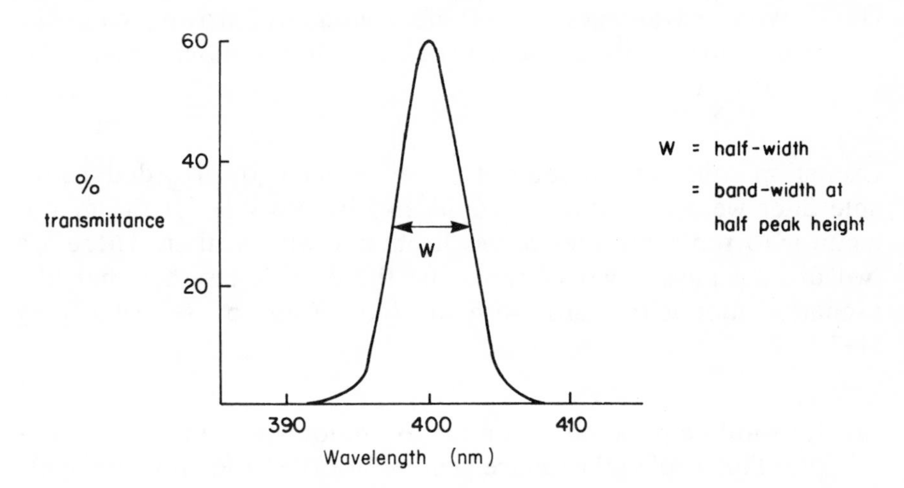

Fig. 2.5d. *A typical transmission curve of an interference filter*

monochromators in terms of light-gathering power and can sometimes out-perform the more expensive instruments in detecting low levels of fluorescence.

Π The use of filters with a mercury source makes particularly good sense. Why? Which monochromator would be replaced?

Mercury sources emit high intensity radiation at only a very limited number of wavelengths (see Section 2.2.2). Interference filters can be made to have their peak transmission at any desired wavelength and filters to isolate the principal mercury lines are readily available. The filter would replace the excitation monochromator.

Π What disadvantage would such an instrument have compared with a dual monochromator instrument?

It would be possible to run the emission spectrum of a compound but not its excitation spectrum. Excitation would be restricted to the wavelengths of the six principal lines of the mercury spectrum. In practice this would still allow most compounds to be excited because excitation bands are generally very broad but in some cases the efficiency would be very poor.

Π What advantages would the xenon arc provide over the mercury lamp as the source in a filter/monochromator instrument?

Excitation could be carried out at any wavelength provided a suitable filter were available. It would also be possible for the manufacturer to replace either monochromator with a filter. There are two distinct advantages of replacing the emission rather than the excitation monochromator with the filter. Can you see what they are?

(*a*) It would be possible to run the excitation spectrum of the sample. This is of rather more general use than the emission spectrum because it is identical with the uv absorption spectrum and can sometimes be used to identify a compound at concentrations too low to record the absorption spectrum.

(*b*) It is quite possible to measure fluorescence with no filter at all on the emission side if a suitable filter is not available. Adequate selectivity can often be attained by careful selection of the excitation wavelength.

2.5.3. Filter Instruments

For many purposes in analytical chemistry it is not necessary to run the full excitation or emission spectrum. In a routine quantitative analysis both the excitation and emission wavelengths remain fixed at the values previously determined when the method was first worked out. In this situation both monochromators can be replaced by suitable band-pass filters to transmit the appropriate wavelengths.

Π What are the main disadvantages of simple filter fluorimeters?

(*a*) Separate filters are required for each analysis to be undertaken.

(*b*) Excitation and emission spectra cannot be obtained. This deprives the operator of the opportunity to check whether there is any unforeseen interference present during a particular analysis.

Π What are their chief advantages?

(*i*) Low cost
(*ii*) Simplicity – very little to go wrong!
(*iii*) Easily portable, require little bench space.

Auxiliary filters are also provided in dual monochromator instruments for specific purposes. These are not band-pass filters of the type we have been discussing but cut-off filters which transmit all radiation above a specified wavelength and absorb completely at lower wavelengths. These are used to remove

(*a*) second (and higher) order grating radiation

(*b*) the scattered exciting radiation before it enters the emission monochromator. This helps to reduce stray light.

The transmission curve of a typical cut-off filter is shown in Fig. 2.5e.

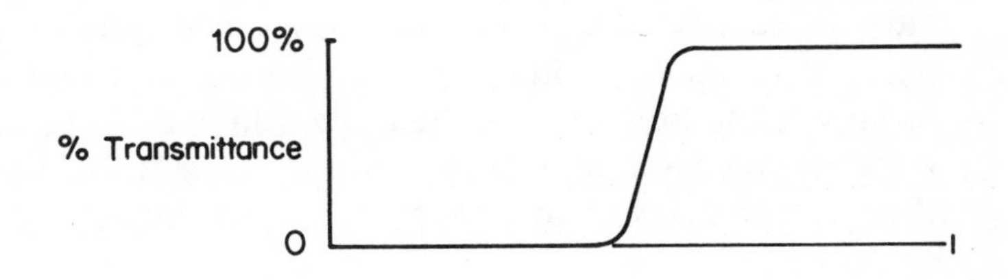

Fig. 2.5e. *Transmission curve of a typical cut-off filter*

Π Add a wavelength scale to the diagram to depict a suitable filter for reducing the intensity of exciting radiation at 435 nm when recording an emission spectrum extending from 460 to 520 nm. Where would this filter be placed in the optical system?

The scale is shown in Fig. 2.5f. The filter would be placed between the sample and the emission monochromator.

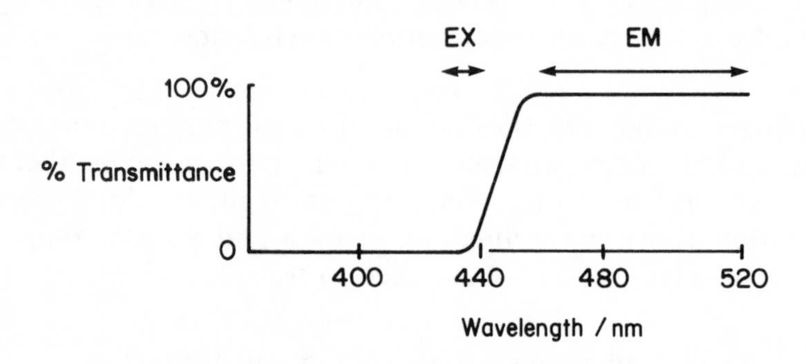

Fig. 2.5f. *Transmission curve for a cut-off filter (cut-off 450 nm and above)*

Π We referred earlier to the fact that some manufacturers provide an accessory which enables uv spectrophotometers to be used to measure fluorescence. You should now be able to appreciate how this can be done. Make a list of the components required, and draw a simple diagram to show how they are positioned to convert the instrument into a filter fluorimeter.

The components required are:

1. A xenon or mercury lamp of adequate intensity together with its power supply.

2. A cell holder to be mounted in the normal sample position with provision for the sample (in a proper fluorimeter cell) to be irradiated at right-angles to the beam direction.

3. An excitation filter placed between the lamp and the sample.

4. An emission filter placed between the sample and the detector.

The layout is shown in Fig. 2.5g.

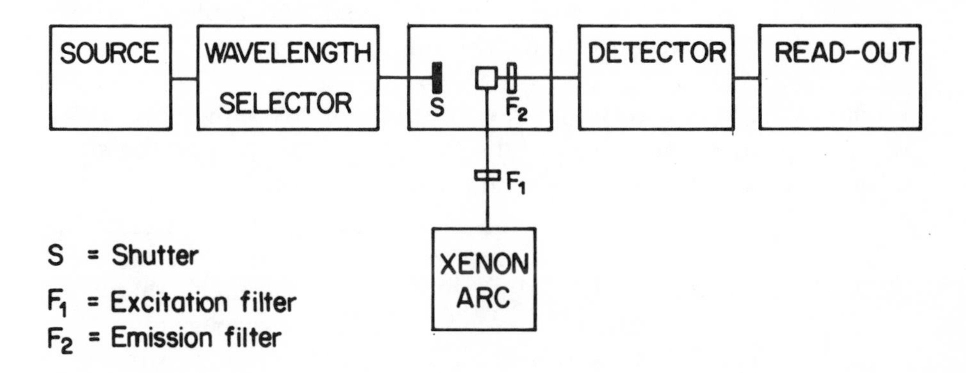

Fig. 2.5g. *Layout of an absorption spectrometer, modified to measure fluorescence*

The shutter is required to cut off radiation from the instrument source only if this cannot be switched off independently. If the spectrometer is a double beam instrument (as it probably will be) it will be necessary to switch it to single beam operation. It will also be necessary to modify the read-out system to record intensity instead of absorbance and possibly to provide a chopper if the detector uses an a.c amplifying system (which it most commonly does).

(These changes parallel exactly the changes necessary to convert an atomic absorption spectrophotometer into a flame emission instrument if you are familiar with that technique.)

The accessory would normally be built round a substantial frame which would slot into the sample position of the instrument to support the xenon arc, filters, chopper and sample.

You might have expected this accessory to make use of the monochromator already present. Why do you think that instrument manufacturers don't favour this?

It is much more difficult and expensive to do it! It would entail placing a second detector at right-angles to the sample beam and connecting it up to the amplifier in place of the original detector. The source unit would have to be removed and replaced with a xenon arc which would require careful alignment.

Finally, it is very likely that the aperture of the monochromator, whilst entirely satisfactory for absorption work, would be too small to provide adequate sensitivity in a spectrofluorimeter.

SAQ 2.5a

If you wanted to buy a filter/monochromator instrument to record the emission spectrum of a sample, which of the monochromators would be replaced by the filter?

SAQ 2.5b

Explain how you would determine the wavelengths of maximum excitation and emission for a compound for which they were not already known. You may assume that the emission monochromator of your dual monochromator spectrofluorimeter has a 'direct light' setting in which the grating acts as a mirror to reflect all the emitted radiation on to the detector.

2.6. CORRECTION OF SPECTRA

We have seen that the intensity of radiation emitted by the source varies considerably with wavelength (Fig. 2.2a). Similarly the output from the detector varies with wavelength for a constant radiation intensity. Furthermore, the energy throughput of a grating monochromator also varies with wavelength, particularly when blazed gratings are employed (which is usually the case). Prism monochromators suffer to a much lesser extent from this defect. Consequently when the fluorescence spectra of a sample are run they are distorted by these variations which are superimposed upon the true spectrum. Hence, although the excitation spectrum should be identical to the absorption spectrum, it is in practice often very markedly different.

Π What quantities are plotted in a true excitation spectrum?

The intensity of fluorescence at a fixed emission wavelength is recorded as the excitation wavelength is altered.

Π How are the variations in source output and detector sensitivity eliminated when recording an absorption spectrum?

By the use of double beam instruments in which the radiation from the source is split into sample and reference beams. The ratio of sample/reference intensities is then used to compute the % transmittance or absorbance which is independent of instrumental variations with wavelength.

Fluorescence spectrometers are inherently single beam instruments and so it is necessary to adopt a different approach to produce a corrected spectrum. Ideally what is required is a material which can be used as a 'quantum counter' to emit radiation whose intensity varies directly with the intensity of the exciting radiation regardless of its wavelength. The compound most commonly used for this purpose is Rhodamine B in solution in glycerol at the relatively high concentration of 5 g dm^{-3}. This is almost totally absorbing over a wide range of the uv region and provides a sample whose fluorescence efficiency is effectively constant (to within 2%) from 350 to 600 nm.

The most efficient way of using this quantum counter is to place a beam-splitter in the beam emerging from the excitation monochromator to direct a small fraction (about 5%) of the radiation on to the solution contained in a suitable cuvette.

The fluorescence emitted by the Rhodamine B is observed at about 615 nm on the tail of the emission band to avoid errors due to re-absorption. A suitable filter is placed in front of a reference photomultiplier for this purpose. The arrangement is frequently built into the excitation monochromator housing as shown below.

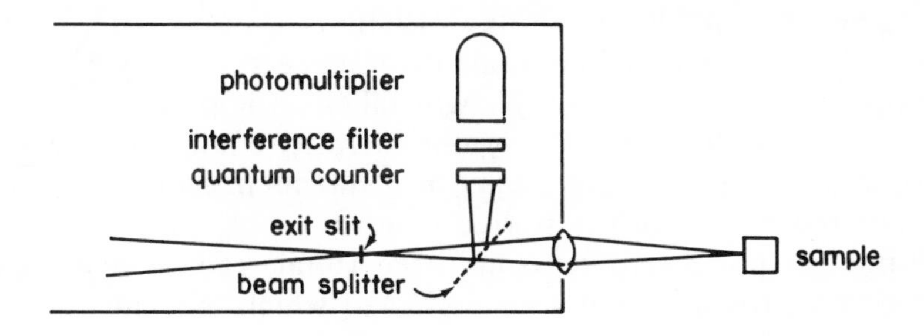

The fluorescence spectrum excitation of the sample can now be corrected automatically by using the signal from the quantum counter photomultiplier as a reference signal in the same way as that from the reference beam in a double beam spectrophotometer. The reference signal is ratioed with the signal from the detector on the emission side of the instrument. This 'on-line' system not only corrects the spectrum for variation of source intensity with wavelength but also for fluctuation of intensity with time. It also takes care of the non-linearity of the energy throughput of the excitation monochromator.

Π Why does the variation of photomultiplier response with wavelength not affect the excitation spectrum?

Because the wavelength of radiation reaching the detector does not change. (Excitation spectra are recorded at fixed emission wavelength.)

An alternative and somewhat less expensive method of correction has become available with modern microprocessor-controlled instruments which have the facility to store and manipulate spectra. Excitation spectra of the sample and the quantum counter (with the emission monochromator set to 615 nm) are run in the normal way and stored separately. The sample spectrum is then divided point by point by the reference spectrum to obtain the corrected spectrum.

Π What disadvantage does this method suffer from compared with the on-line method?

It is subject to error if there is any variation in source intensity with time. This can be very important in practice since a typical fluorimetric analysis at fixed wavelength takes an appreciable time to complete and any variation in source intensity could give rise to serious errors if uncorrected. Many instruments provide a system to monitor the output of the source in which a photocell or photomultiplier is positioned to sample the (total) radiation from the source and provide a reference signal to which the sample signal is ratioed.

Even for fixed wavelength work this is less satisfactory than full online correction since it does not take account of any variation of the intensity/wavelength characteristic with time.

The corrected excitation spectrum is desirable when it is to be compared with an absorption spectrum or when comparing spectra recorded on different instruments. Correction of the emission spectrum is much less important except in more fundamental studies such as the determination of fluorescence efficiencies. It involves calibrating the emission monochromator and detector with a source of known spectral output and is a rather lengthy and tedious business. It is seldom carried out in analytical work.

SAQ 2.6a

What instrumental factors affect the appearance of fluorescence spectra, and make correction necessary when comparing spectra from different instruments? Distinguish between those factors which affect the excitation spectrum and those which affect the emission spectrum.

SAQ 2.6b

When is it necessary to correct fluorescence spectra? For what purposes is correction unnecessary?

SAQ 2.6b

SAQ 2.6c

What is a quantum counter? Explain how it is used to correct excitation spectra.

2.7. INSTRUMENTATION FOR PHOSPHORESCENCE

The only essential practical difference between the processes of phosphorescence and fluorescence is that phosphorescence is of much longer lifetime and therefore easily quenched unless, as we saw in Part 1, the sample is in the solid state. You will not be surprised to find therefore that the instrumentation required for observing phosphorescence is essentially similar to that used for fluorescence. In fact it is standard practice to use a special accessory to permit phosphorescence to be measured with a spectrofluorimeter. This involves changes in the sampling technique and in the procedure for recording the spectra. Try to visualise what these changes might be.

Take the sampling technique first. As we have noted the sample must be in the solid state or the phosphorescence will be quenched. It is of course possible to use a simple accessory for placing a solid sample in the instrument so that the surface is at 45° to the incident beam as shown here.

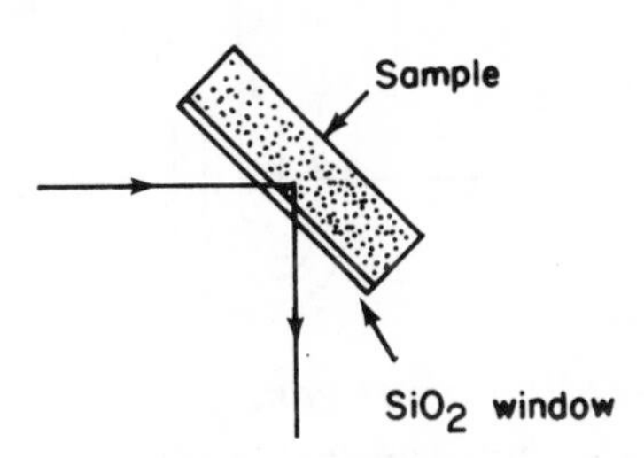

Because of the very high concentration of the neat solid sample, light does not penetrate very far into it and so the phosphorescence (and fluorescence) is observed only from the surface layer. Furthermore it is accompanied by a very high level of direct scattering of the exciting radiation. This would be a serious problem with fluorescence but, because of the longer lifetime of phosphorescence, it can be observed free of interference from scattering (or fluorescence) by techniques which we shall discuss shortly. Consequently, useful results can be obtained for powdered samples and samples absorbed on filter paper and the technique of *room temperature phosphorescence* (RTP) is now an exciting new development in photoluminescence spectroscopy.

For accurate quantitative analysis, however, a solvent system is generally needed to dilute the sample to the concentration levels normally used for fluorescence work (1 mg dm^{-3} or less). This has to have some rather special properties. In particular, it must freeze to give a clear 'glass' so that direct scatter of exciting radiation will be reduced to a minimum. For polar materials ethanol is an excellent solvent, whilst for non-polar samples a mixture of diethyl ether, isopentane and ethanol in the ratio 5 : 5 : 2 respectively has been found to be most effective. The solution is usually frozen in liquid nitrogen contained in a Dewar vessel with a non-silvered extension at its bottom end. The solution is contained in a narrow silica tube about 2 mm in diameter to ensure that it freezes to a glass without cracking which would be much more likely if a large sample were used. This part of the accessory is shown in Fig. 2.7a.

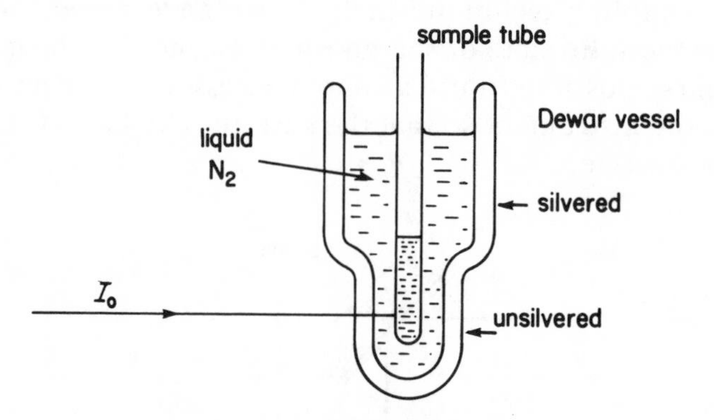

Fig. 2.7a. *Sampling arrangement for low temperature phosphorescence*

The need to freeze samples in liquid nitrogen has been a considerable disincentive to analysts in the past and, in consequence, phosphorescence has been very little used in everyday analysis. An alternative method of protecting triplet state molecules from quenching has recently been developed which enables phosphorescence to be observed with samples in the liquid phase at room temperature. This could well lead to a much wider use of phosphorescence in the future. The technique is described later in Part 5.

Once the sample is frozen, both its fluorescence and phosphorescence emission spectra could be observed in the normal way since the phosphorescence will be at much longer wavelength than the fluorescence. However, better results are obtained by separating the phosphorescence from the fluorescence signal by taking advantage of the longer lifetime of the phosphorescence. If the incident radiation is cut off, fluorescence ceases almost immediately together with direct scattering. Phosphorescence however continues and can be recorded free from interference from fluorescence or scattered radiation for some time afterwards (tens of milliseconds at least). This principle can be applied by using either a mechanical shutter system or a pulsed source with an electronic timing mechanism which records the detector output only during the 'dark periods' – a technique known as *time resolution* or *gating*.

Two shutter systems are commonly employed

(*i*) a revolving can surrounding the sample with a small window which admits a short burst of exciting radiation to the sample when it is opposite to the excitation monochromator and a short burst of phosphorescence when it is opposite the emission monochromator as shown in Fig. 2.7b. The delay between

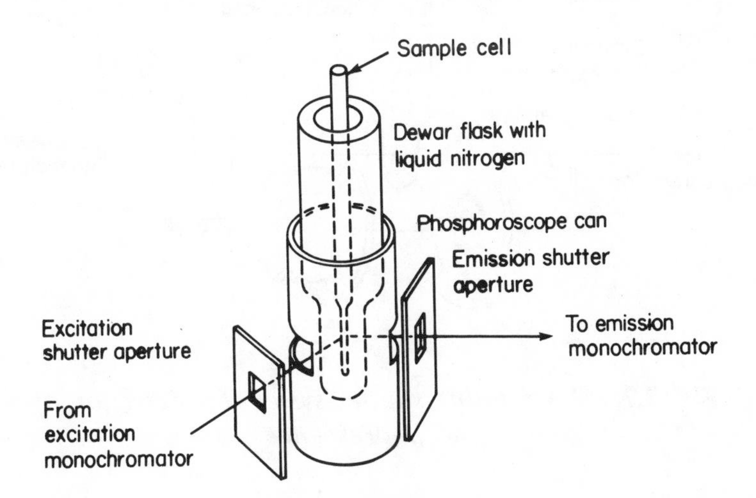

Fig. 2.7b. *A revolving can shutter system for detection of phosphorescence*

excitation and emission is determined by the speed of rotation (typically >1000 rpm) and the length of each observation by the size of the window. Shorter delay times can be obtained by using a number of equally spaced windows which also helps to balance the rotating system and reduce vibration.

You will notice that with this technique the incident beam and the emitted beam are in the normal 90° configuration.

(*ii*) a rotating shutter having two discs, one on each side of the sample, with opposing 90° sectors cut out of each disc. With this device the phosphorescence is recorded as soon as the excitation ceases (though a delay can be achieved by reducing the angle cut out to less than 90°) – twice in each revolution as shown in Fig. 2.7c. The duration of each burst of radiation (and the delay between them if any) is determined by the speed of rotation. Unlike the rotating can, the shutter system requires 180° geometry.

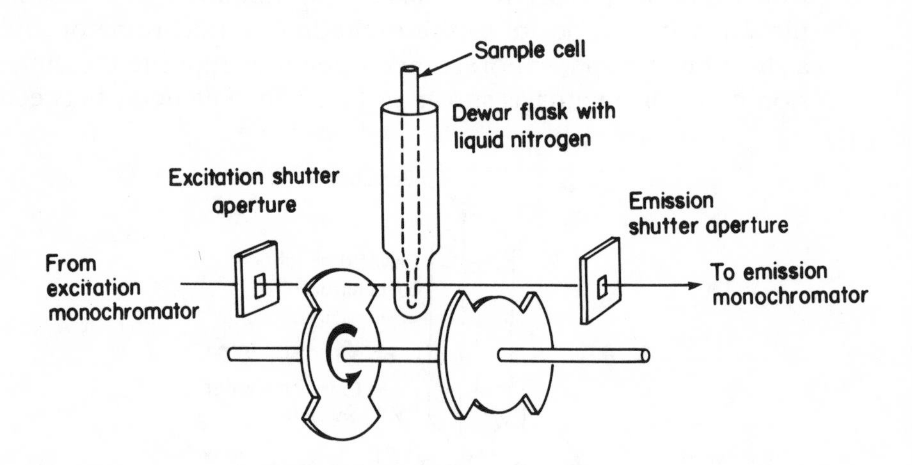

Fig. 2.7c. *A rotating shutter system for detection of phosphorescence*

Electronic gating has come to the fore in recent years with the increasing use of microprocessors in analytical instruments and the current trend towards pulsed sources for fluorescence work. The advantage of the pulsed xenon arc over the continuous output arc is that a far higher intensity can be produced during the pulse without overheating the lamp since the thermal energy is dissipated in the dark periods between the pulses. The intensity versus time profile of the fluorescence and the scattered radiation follow closely that of the lamp pulses and the overall (integrated) intensity is considerably higher than that obtained with a continuous source. The intensity/time curve of the emission from the sample then appears as in Fig. 2.7d.

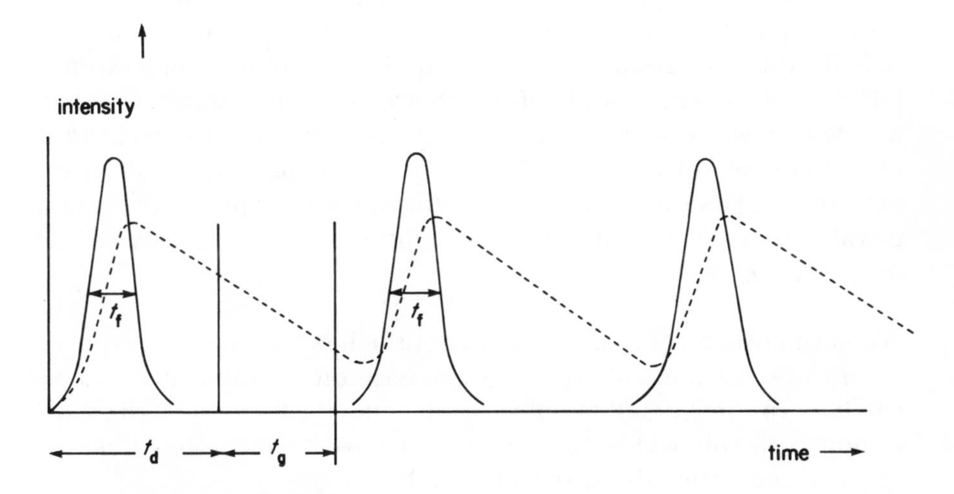

Fig. 2.7d. *Timing sequence for the detection of phosphorescence in a spectrometer employing a pulsed radiation source*

The full line in this diagram refers to fluorescence emission and scattering, the broken line to phosphorescence. It is a routine matter with a microprocessor and an accurate clock system to extract the signal from the phosphorescence in the dark periods between the pulses. Both the duration of the observation (the width of the gate, t_g) and the delay from the start of the pulse, t_d, can be varied. The half width of the pulse (typically about 10 μs) and the pulse repetition rate (about 100 flashes per sec) are not normally varied though, in principle, there is no reason why they should not be.

The pulsed source system allows for a much greater control and flexibility in the time resolution and for observation of phosphorescence with much shorter lifetime than is possible with the mechanical system, whose performance is limited by the maximum speed of rotation available. A modern instrument of this type can be used not only for the more routine analytical application of phosphorescence but also for a detailed study of the lifetime of excited states. We shall not be concerned with the more academic aspects of the technique in this course but the control over the timing parameters does extend the range of analytical applications, for example by making it possible to analyse mixtures of phosphorescent compounds having different lifetimes.

We have now completed our survey of the instrumentation required for the observation and measurement of photoluminescence and we can look in some detail at the ways in which the techniques are used in analysis. This will be the subject of Part 3 which will illustrate many of the points discussed in Part 2.

SAQ 2.7a

How do the characteristics of the radiation entering the monochromator in a phosphorescence instrument equipped with a pulsed source and electronic timing mechanism differ from that in an instrument with a continuous source and a rotating can or shutter system?

SAQ 2.7a

SAQ 2.7b

In what respects does the performance of the detector and the monochromator used for phosphorescence work need to be superior to the minimum required for fluorescence work?

SAQ 2.7b

Summary

The instrumentation for observing photoluminescence has many similarities to that used for measuring absorption in the uv region of the spectrum. The sample is excited by radiation from a high power xenon arc or mercury lamp at a wavelength determined by a wavelength selector. The emitted radiation is usually observed at right-angles to the incident beam through a second wavelength selector. A detector and read-out device completes the system. The sample is contained in a cell similar to that used for absorbance measurement except that in most cases all four sides are polished.

Several types of fluorescence instrument are available using various combinations of xenon or mercury source, grating monochromators or filters, photomultiplier or solid state detector and analogue or digital meter on chart recorder read-out. They are generally classified in terms of the choice of wavelength selectors.

The most versatile and expensive instruments have two grating monochromators of high aperture. These can be used to record both excitation and emission spectra or to carry out quantitative analysis at fixed wavelength. The combination of a band-pass filter with a grating monochromator provides a cheaper instrument which is capable of recording only the emission or the excitation spectrum. The mercury lamp with a set of excitation filters to isolate the principal mercury lines is sometimes used in this type of instrument which is also suitable for fixed wavelength work. For routine quantitative analysis at fixed wavelength simple filter instruments are quite adequate. Another type of filter with a sharp cut-off just above the excitation wavelength is often provided on grating instruments to reduce stray light problems. Some manufacturers of uv absorption instruments supply accessories to enable them to measure fluorescence. These are usually simple filter arrangements which do not make use of the existing monochromator.

Fluorescence is inherently a 'single beam' technique so fluorescence spectra are distorted by variations of source output and detector sensitivity with wavelength. In order to compare spectra from different instruments or, more particularly, excitation spectra with absorption spectra it is necessary to correct the spectrum. The excitation spectrum is corrected by ratioing the emission signal to a reference signal generated by a quantum counter irradiated by a small sample of the beam emerging from the excitation monochromator. Correction of the emission spectrum is possible in principle but seldom applied in practice for analytical work. It is, however, common practice to monitor the undispersed radiation from the source to provide a reference signal to which the analytical signal can be ratioed to eliminate variations in fluorescence intensity due to lamp fluctuations. Modern instruments are commonly controlled by microprocessors which are also used to process the output. This also provides an alternative way of correcting spectra.

The instrument used to observe phosphorescence is essentially a fluorescence instrument provided with a system to interrupt the incident radiation so that the long-lived phosphorescence can be recorded during the dark periods. Earlier instruments used a mechanical chopping system with a continuous source. Modern microprocessor-controlled instruments use a pulsed source with electronic gating on the detector output.

In order to avoid quenching the phosphorescence, the sample is normally examined in the solid state, either as a powder or on a filter paper at room temperature, or frozen to a 'glass' in a suitable solvent in liquid nitrogen. The latter gives better results, particularly in quantitative analysis, but the former is more convenient and more widely used for qualitative and rough quantitative work. Other techniques are now available to enable phosphorescence to be observed at room temperature with liquid samples.

Objectives

You should now be able to:

- recognise how instruments used for fluorescence are related to those used for uv/visible absorption measurements;

- describe the radiation sources used in fluorescence instruments, their characteristics, and be aware of the hazards associated with them;

- explain the advantages of 90° geometry in fluorescence instruments;

- compare the cells used for fluorescence with those used in absorption spectroscopy and appreciate the need to take appropriate precautions to preserve their optical quality;

- draw a block diagram of a double monochromator spectrofluorimeter and appreciate the requirements on the monochromator in fluorescence instruments;

- explain how the full excitation and emission spectra of a sample are recorded;

- explain the scope and limitations of filter monochromator and double filter instruments;

- recognise the need for the correction of excitation and emission spectra and explain how this is done;

- describe the modifications necessary to a spectrofluorimeter to enable it to be used for phosphorescence measurements;

- describe the techniques of mechanical chopping and electronic gating in phosphorimeters and explain their relative merits.

3. Quantitative Fluorimetry

3.1. INTRODUCTION

You should by now have a reasonable understanding of the principles of fluorescence spectroscopy and a fair appreciation of the practical aspects of the technique. The time has now come to discover how the technique is used in analytical chemistry. By far the most important application of fluorescence is in quantitative analysis. In this respect its capabilities are quite similar to those of uv/visible absorption spectroscopy. This should not surprise you since both techniques depend upon electronic transitions in molecules. Electronic spectra are of limited value in the identification of 'unknown' compounds because they usually consist largely of broad featureless bands which are not sufficiently distinct. This is in marked contrast to techniques such as vibrational (ir) and nuclear magnetic resonance (nmr) spectroscopy although electronic spectroscopy (both uv absorption and fluorescence) can sometimes provide useful confirmatory evidence in cases which have not yielded to examination by the other techniques.

On the other hand, electronic absorption spectra have for many years been used for quantitative analysis due mainly to the practical advantages of the technique. Most important of these are its high sensitivity and the fact that common solvents such as hexane, ethanol and, particularly, water are completely transparent to visible and uv radiation above 210 nm. This makes it possible to use the technique for trace analysis of a wide range of analytes in the natural environment which accounts for a very large slice of modern analytical chemistry. Both metals and organic materials can be determined though much of the metals analysis is now carried out by the techniques of atomic spectroscopy such as atomic absorption and plasma emission.

Once the sample is in solution, the measurement of absorbance is very convenient and the use of glass or silica cells, which can be manufactured to close dimensional tolerances, makes it possible to achieve relatively high precision in the analysis. Finally, the cost of the minimum equipment required is much lower than for many other techniques.

As we shall see, fluorescence measurement is equally convenient and offers the prospect of even greater sensitivity. It also has the potential for greater specificity than absorption though this is not always realised in practice. On the other hand, the precision of fluorescence measurements is not as good – though at very low trace levels this is often of no great consequence.

3.2. CALIBRATION GRAPHS

As with all other instrumental methods of analysis, the quantitative application of absorption and fluorescence spectroscopy depends upon the measurement of an 'analytical signal' which varies with the concentration of the analyte. The signal obtained for the sample being analysed is converted to a concentration value by means of a 'calibration graph' prepared by measuring the signals for a number of 'standard solutions' containing known quantities of the analyte.

Quantitative analysis by absorption spectroscopy therefore involves the measurement of the absorbance of the sample contained in a cell manufactured to high precision, usually 1 cm square, subtraction of the absorbance of the blank contained in a similar cell ('matched pairs' of cells with identical absorption characteristics over a wide range of wavelength are standard items of equipment) and conversion to concentration using a calibration graph. Calibration curves are generally linear though occasionally so-called 'deviations from Beer's Law' result in a curved graph.

Now let us see how fluorescence measurements are related to the concentration of the fluorescent species. We have already discussed the nature of fluorescence and the experimental conditions necessary for its observation in Parts 1 and 2 so you should be able to predict what factors will be involved.

Π What is the origin the analytical signal in a fluorescence spectrometer?

The fluorescence emitted by the sample.

Π What instrumental factors will affect the signal?

The chief factors are the intensity of the incident radiation and the efficiency of the optical system. The sensitivity of the detector is also relevant since this determines the magnitude of the photo-current produced as a measure of the analytical signal.

Π What parameter in the fluorescence process is equivalent to the absorptivity in absorption?

The fluorescence efficiency of the fluorescent species, ϕ_f.

You might reasonably expect that the intensity of fluorescence will be directly proportional to the intensity of the incident radiation and to the fluorescence efficiency. It is also likely that optical factors for a given instrument will be constant for a particular sample configuration. The other important factor is that the intensity of fluorescence will depend upon the amount of radiation absorbed and this is the factor which is determined by the concentration of the fluorescent species.

Let's try to quantify these factors for 90° sample geometry, first reminding ourselves of the experimental arrangement by means of the simple diagram we used in Part 2:

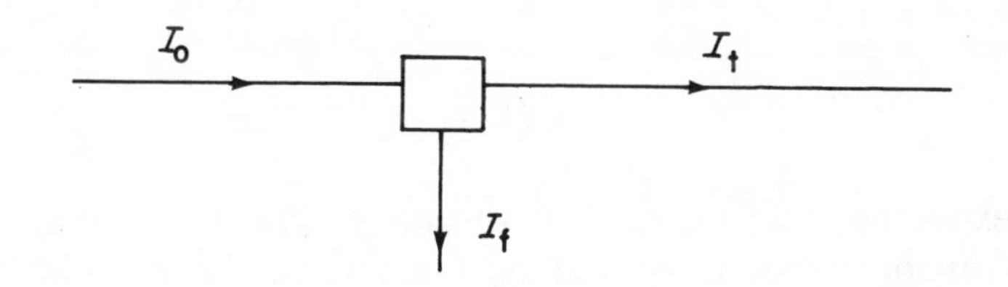

Π We saw in Part 1 that the definition of fluorescence efficiency involves I_f. Can you remember what it is?

The fluorescence efficiency, ϕ_f, is defined as the fraction of the incident radiation which is re-emitted as fluorescence.

$$\phi_f = \frac{\text{no. of photons emitted}}{\text{no. of photons absorbed}} = \frac{\text{intensity of fluorescence}}{\text{intensity of absorption}} = \frac{I_f}{I_a}$$

We can rearrange this definition to give an expression for the intensity of fluorescence, I_f

$$I_f = \phi_f I_a \qquad (3.1)$$

I_a is simply the intensity of radiation not transmitted, $I_o - I_t$.

Substituting this in Eq. 3.1 gives

$$I_f = \phi_f (I_o - I_t) \qquad (3.2)$$

or

$$I_f = \phi_f I_o(1 - I_t/I_o) \qquad (3.3)$$

Π How does the term I_t/I_o help us to introduce the concentration into our expression?

This is the transmittance of the sample which is related to the concentration by the Beer–Lambert law. We have to insert the definition of the absorbance into the simple expression $A = \epsilon cd$ to get $\log(I_o/I_t) = \epsilon cd$. You will see shortly that we really need the version in natural logs, $\ln(I_o/I_t) = kcd$, which becomes, when expressed in its exponential form, $I_t/I_o = e^{-kcd}$

hence

$$I_f = \phi_f I_o(1 - e^{-kcd}) \quad (3.4)$$

As we expected, this expression shows clearly that the analytical signal is proportional to I_o and ϕ_f. Unfortunately the relationship to the concentration is not linear. Why not?

The concentration appears as an exponential function. This is not very convenient so let's see what we can do about it. The standard procedure for getting rid of awkward exponential functions is to use the fact that they can be expressed as a power series:

$$e^x = 1 + x/1! + x^2/2! + x^3/3! + x^4/4! + \ldots \text{ etc} \quad (3.5)$$

In our case, we shall want the expression for e^{-x} which involves changing the signs of the odd powers:

$$e^{-x} = 1 - x/1! + x^2/2! - x^{/}3! + x^4/4! \ldots \text{ etc} \quad (3.6)$$

(The terms in the denominators are called 'factorials' where

$$3! = 3 \times 2 \times 1 \qquad 4! = 4 \times 3 \times 2 \times 1 \text{etc.})$$

The next step is to note that, if x is very small, we can ignore the terms in x^2, x^3 and higher powers without serious loss of accuracy and use the simple substitution $e^{-x} = 1 - x$. Carry out this substitution on Eq. 3.4.

Hey presto! We now have an expression for I_f which is linear in concentration:

$$I_f = \phi_f I_o kcd \quad (3.7)$$

The expression in the bracket in Eq. 3.4 $(1 - e^{-kcd})$, becomes $(1 - 1 + kcd) = kcd$.

The final step is to reinstate the molar absorptivity, ϵ, in our expression in place of k which appeared on the scene when we used the exponential form of the Beer–Lambert expression to get Eq. 3.4. All we have to do is to remember that

$$\ln(1/T) = 2.303 \log(1/T)$$

so that

$$kcd = 2.303\, \epsilon cd$$

and hence

$$k = 2.303\, \epsilon$$

This gives us our working version of the fluorescence formula

$$I_f = 2.303\, \phi_f I_0 \epsilon cd \qquad (3.8)$$

The quantity I_f refers to the total fluorescence intensity in all directions and so the right-hand side of Eq. 3.8 will eventually have to be multiplied by a suitable factor to take into account the actual fraction of fluorescent radiation collected by the emission monochromator.

In order to get an expression for I_f which is linear in concentration we had to make the assumption that kcd is very small so that we could ignore the higher terms in the expansion of e^{-kcd}. Let's have a closer look at the full expression for I_f to see what 'very small' means in practice. If we include the terms in x^2 and x^3 the expression for I_f becomes

$$I_f = \phi_f I_0 \{2.303\, \epsilon cd - (2.303\, \epsilon cd)^2/2 + (2.303\, \epsilon cd)^3/6\} \qquad (3.9)$$

Π Before we investigate the magnitude of the effect of the higher term on the linearity of the graph, which way would you expect it to curve – towards the ordinate (intensity) or the abscissa (concentration) axis?

Towards the abscissa axis.

The square term has a negative sign and so will reduce the value of I_f to a greater extent as the concentration increases. The cubic term has a positive sign and so will reduce the curvature though its effect will be negligible until much higher concentrations are reached. This result is similar to that observed in absorption spectroscopy when there is a negative deviation from Beer's law (the most common type of deviation) but the causes are of course completely different.

Let's suppose that we use a standard 1 cm cell so that $d = 1$ cm. Let's further suppose that we have a fluorescent compound whose molar absorptivity is 10 000 $dm^3\ mol^{-1}\ cm^{-1}$ (1000 $m^2\ mol^{-1}$) – a typical value for a compound such as anthracene.

Note that the value of the molar absorptivity given in brackets is the value in SI units. Unfortunately this has not caught on in the real world of organic and analytical chemistry and in practice the value is almost always quoted in $dm^3\ mol^{-1}\ cm^{-1}$, or, more often, as $l\ mol^{-1}\ cm^{-1}$. This is less cumbersome and quite permissible, since, even in the SI system, the litre is the accepted alternative name for the cubic decimetre. Unfortunately, the molar absorptivity is often quoted as a pure number in the literature, the unit '$l\ mol^{-1}\ cm^{-1}$' being understood. In this Unit we shall use $dm^3\ mol^{-1}\ cm^{-1}$. Since ϵ does not normally occur in expressions involving other quantities which are expressed in SI units this causes no problems in other areas of chemistry.

The graph obtained by inserting these values in Eq. 3.9 for the concentration range 0–30 μmol dm^{-3} is shown in Fig. 3.2b together with the graph obtained by omitting the cubic term.

Π At what concentration does

(*a*) the graph depart from linearity; and (*b*) the cubic term become significant?

(*a*) below 4 μmol dm^{-3} and (*b*) about 14 μmol dm^{-3}

Π It is rather difficult to estimate the limit of linearity from Fig. 3.2b so the range 0–5 μmol dm^{-3} is redrawn on a larger scale in Fig. 3.2c. What do you now make the concentration at which the graph departs from linearity?

I make it 2.4 μmol dm^{-3}.

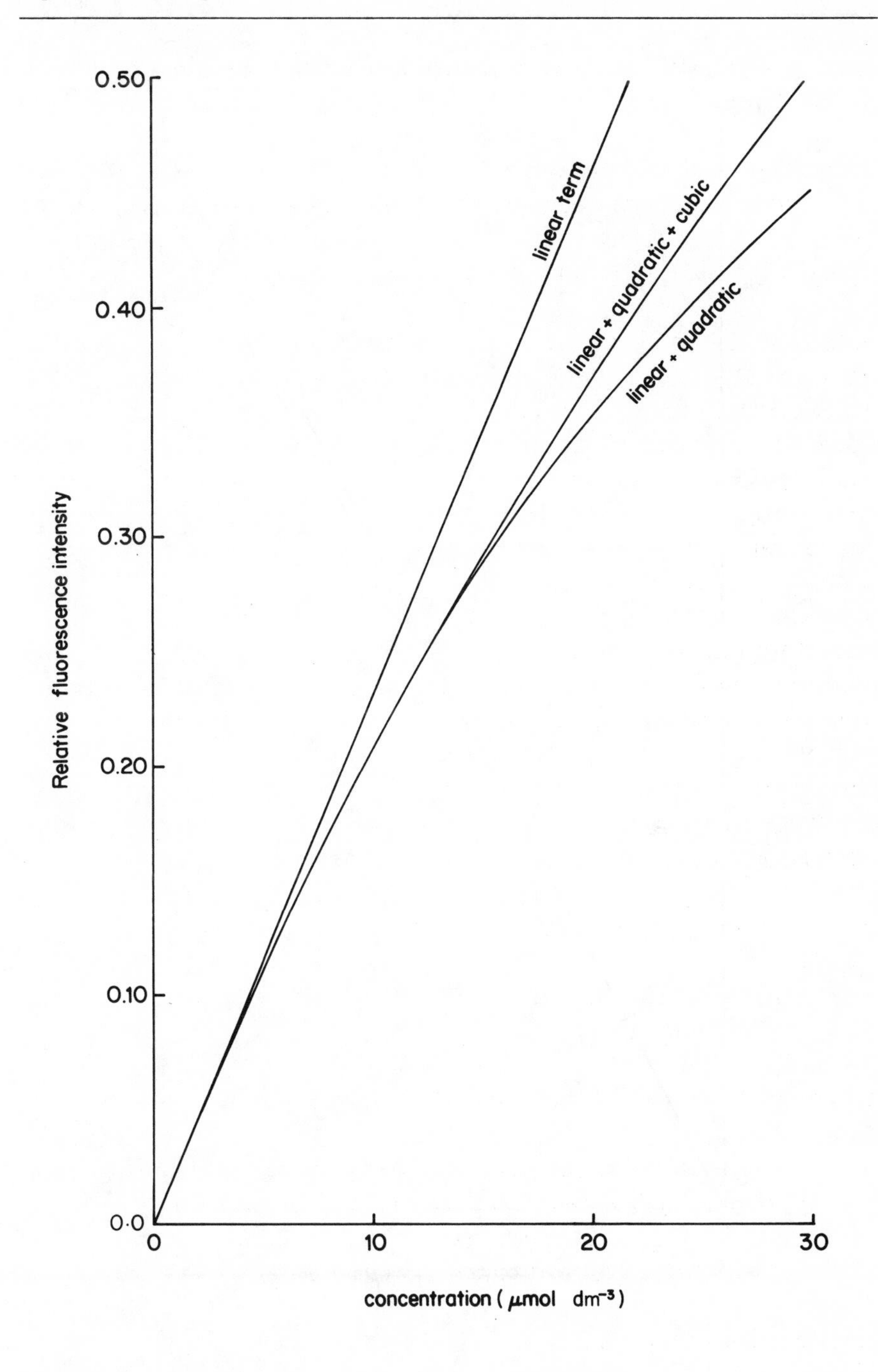

Fig. 3.2b. *Alternative calibration curves for fluorimetric analysis*

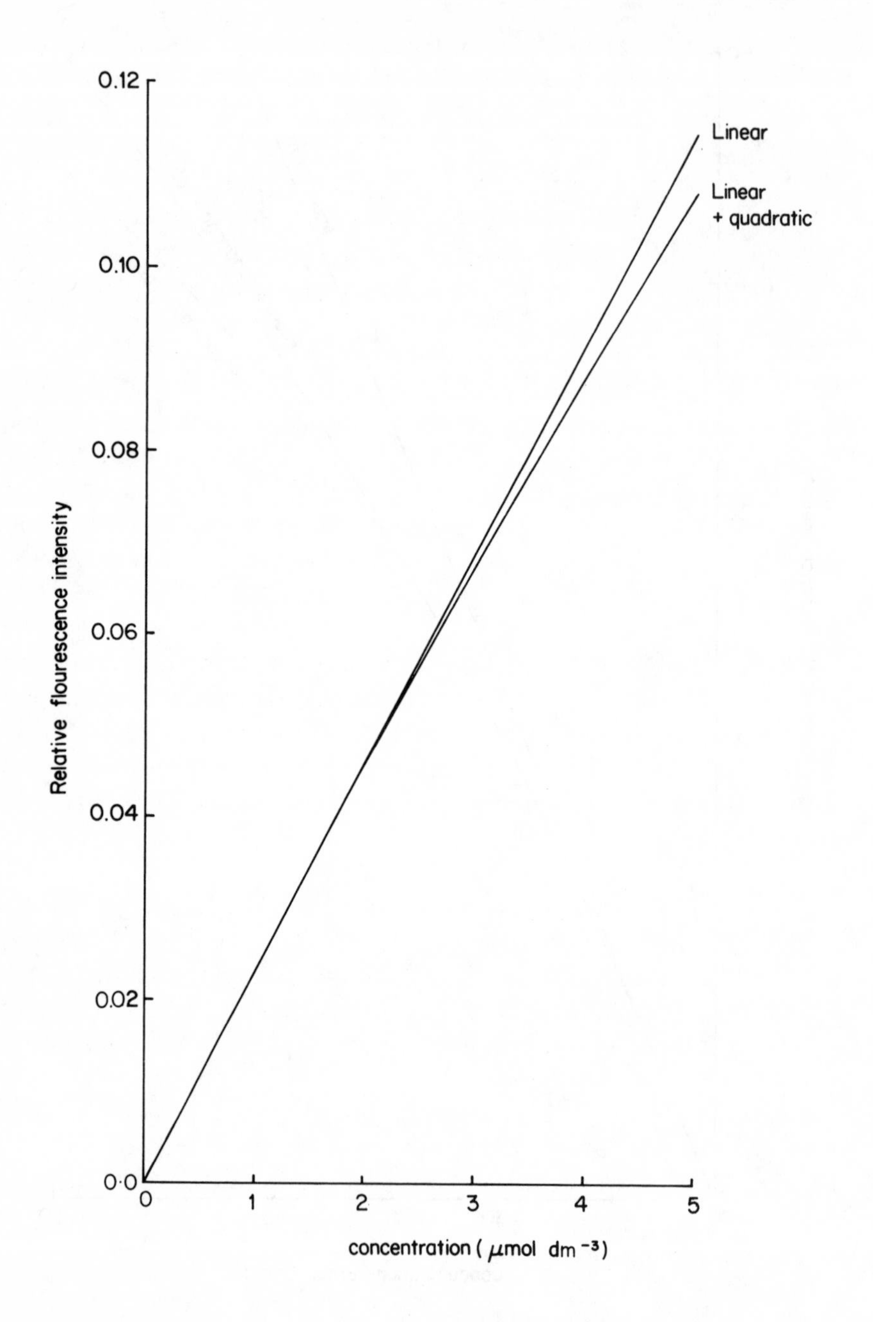

Fig. 3.2c. *Alternative calibration curves for fluorimetric analysis*

Π If the compound has a relative molar mass of 300, what does this concentration become in mg dm^{-3}?

0.720 mg dm^{-3}.

Thus for a typical fluorescent species of relative molar mass 300 and absorptivity 10 000 dm^3 mol^{-1} cm^{-1} (1000 m^2 mol^{-1}) we can expect the calibration to be linear provided we work with very dilute solutions below 2.4 μmol dm^{-3}.

The question now arises, can we actually observe the fluorescence at this concentration? This of course depends upon the sensitivity of our instrument which in turn depends upon I_0, and the efficiency of the emission optics and the detector, which we have no means of estimating. Fortunately it turns out that for compounds with reasonable fluorescence efficiences ($\phi_f > 0.1$) this concentration is several orders of magnitude above the detection limit.

Let's see how this concentration compares with the detection limit in absorption spectroscopy. Since absorption measurements depend upon the value of the absorbance which involves the measurement of a ratio, instrumental factors cancel out and the limit of detection is that concentration which gives an absorbance equal to twice the noise level.

Π If this minimum absorbance value is 0.002 (not unreasonable for a high quality instrument) calculate the limit of detection of the compound we have been discussing, in mg dm^{-3} using a 1 cm cell. How does this compare with the concentration at the upper limit of linearity in the fluorescence curve?

0.060 mg dm^{-3}

This is a straightforward application of the Beer–Lambert Law:

$$\begin{aligned} A &= \epsilon c d \\ 0.002 &= 10\,000 \times c \times 1.0 \\ c &= 0.002/10\,000 \\ &= 0.200 \text{ mol dm}^{-3} \end{aligned}$$

$$= 0.200 \times 300 \text{ g dm}^{-3}$$
$$= 0.200 \times 0.300 \text{ mg dm}^{-3}$$
$$= 0.060 \text{ mg dm}^{-3}$$

The upper limit for linearity of the fluorescence calibration is thus only about 10 times larger than the uv/visible detection limit. The fluorescence detection limit is probably some 1000 times lower.

Note that the ordinate plotted on our graphs is the *relative* intensity of the fluorescence as expressed by Eq. 3.9. The numerical value is obtained by setting I_0 and ϕ_f equal to 1.0 which is permissible for a given compound measured on the same instrument at fixed wavelength. The value of the ordinate at the upper limit of the linear range is the highest value at which I_f can truly be represented by 2.303 ϵcd and is useful since it is independent of the compound concerned.

Π What is this value for the calibration graphs given in Fig. 3.2b?

A reasonable estimate is 0.05 – we can't really be very precise in assessing where the curves actually diverge.

To calculate the highest concentration we can use to get a linear calibration for a particular compound, we simply insert the appropriate value of ϵ and, if we want the concentration in w/v units, the relative molar mass into this expression and equate it to 0.05. Calculate this concentration in mg dm^{-3} for quinine for which ϵ = 4540 dm^3 mol^{-1} cm^{-1} (454 m^2 mol^{-1}) and M_r = 324.

1.5 mg l^{-1}.
2.303 ϵcd = 0.05.

Since d = 1 cm (always assume this unless you are told otherwise)

$$\epsilon c = 0.05/2.303$$
$$= 0.0217$$
$$\text{hence } c = 0.0217/4540 \text{ mol dm}^{-3}$$
$$= 0.0217 \times 324/4540 \text{ g dm}^{-3}$$
$$= 21.7 \times 324/4540 \text{ mg l}^{-1}$$
$$= 1.5 \text{ mg dm}^{-3}.$$

Π What is the absorbance value corresponding to the concentration at the top of the linear range of the fluorescence graph?

The concentration at the top of the linear range corresponds to absorbance of 0.022 since $A = \epsilon cd$ and $2.303\ \epsilon cd = 0.05$.

As a 'rule of thumb' you should remember that, in order to get a linear calibration graph, the absorbance of the top standard should not exceed 0.02.

3.2.3. Dynamic Range

In practice, whilst we should prefer our calibration graph to be linear, it is quite possible to operate if it is curved. This does lead to some loss of accuracy and sensitivity in the sense that the change in intensity is less for a given change in concentration at higher concentrations. However, it does allow the operator to cover a wider concentration range and avoid the inconvenience and possible errors in diluting the sample down to the linear range. The 'dynamic range' of any analytical technique is the highest measurable concentration divided by the lowest concentration measurable under 'routine' conditions (10 times the detection limit is a reasonable figure). For absorption techniques in general this ratio is limited to about 100 which requires accurate measurement over an absorbance range of 0.020 to 2.00. This is possible only with modern, high quality instruments. Fluorescence (in common with other emission techniques) has a dynamic range of at least 10 000. Modern instruments also have the facility to correct the curvature of a curved calibration and so improve the overall accuracy.

3.2.4. Concentration Units

Although the mol dm^{-3} and its subdivisions are the natural and fundamental units for spectroscopic work (since the absorbance and the fluorescence intensity are determined by the number of molecules per unit volume) the weight/volume unit is always used in practice to specify toxic limits for pollutants. Consequently, practising analysts have adopted this convention for all analytical work and normally use mg dm^{-3} or μg dm^{-3} as the unit in all circumstances involving very low concentrations.

Unfortunately there is a widespread use of parts per million (ppm) as a w/v concentration unit though strictly this is a weight/weight quantity. Because of the extreme sensitivity of fluorescence, concentrations of well below '1 ppm' are commonly encountered and the unit 'parts per billion' (ppb) has been introduced to avoid very small numerical values. Unfortunately, in America (and in the world of high finance generally) the billion is interpreted as 1000 million (10^9) whereas its original definition and still commonly understood meaning in Great Britain and Europe is a million million (10^{12}). The unit ppb is therefore ambiguous and you should avoid using it. Similarly the 'part per trillion' (ppt) can mean 1 in 10^{12} or 1 in 10^{18} depending on which side of the Atlantic you use it! Furthermore, 'ppt' is sometimes used as the abbreviation for 'parts per thousand' which adds to the confusion. In general, the weight/volume unit is to be preferred in all circumstances using an appropriate weight unit to give convenient numerical values – mg dm^{-3}, μg dm^{-3}, ng dm^{-3} etc.

Π As a competent analyst you should be able to interconvert concentration units quite freely. See if you can express 10 ppm in mg dm^{-3}, μg cm^{-3} and ng cm^{-3}.

10 ppm = 10 mg dm^{-3} = 10 μg cm^{-3} = 10 000 ng cm^{-3}

10 ppm = 10 mg in 1 000 000 mg = 10 mg in 1000 g = 10 mg dm^{-3} since a litre (1 dm^3) of water or very dilute solution weighs 1000 g.

It is this dependence on the density which is the major objection to the use of ppm for solution work. A litre of sea-water weighs about 1030 g so trace elements present at a level of 10 ppm in fresh water would have a concentration of 9.7 ppm in sea water though its concentration in mg dm^{-3} would be the same in both cases (10 mg dm^{-3}). This difference, though small, is greater than the experimental error.

SAQ 3.2a

Calculate the limiting concentration for linearity in mg dm^{-3} for the following compounds (using 1 cm cells):

		M_r	$\epsilon/dm^3\ mol^{-1}\ cm^{-1}$
(*i*)	β-carotene	534	122 000
(*ii*)	Naphthalene	128	5 600
(*iii*)	Benzene	78	200
(*iv*)	Chrysene	228	139 000

SAQ 3.2b

Identify the phrase which correctly completes the following sentence:

The chief objection to the use of the term 'parts per million' for solutions of low concentration is that

(*i*) it gives unreasonably high numerical values.

(*ii*) it is only valid for solutions of density 1.000 g cm^{-3}.

(*iii*) the million has a different meaning in different countries.

(*iv*) it is not possible to relate it to concentrations in w/v units even when the density is 1.000 g cm^{-3}.

SAQ 3.2c

What additional objections are there to the unit 'parts per trillion' and its abbreviation 'ppt' besides its dependence on density?

SAQ 3.2d

In a particular fluorimetric analysis the limit of detection was 0.35 ng cm^{-3} and the upper limit of concentration was 50 mg dm^{-3}. What is the dynamic range of this analysis?

(*i*) 140 000
(*ii*) 14 000
(*iii*) 1400
(*iv*) 14 285
(*v*) 0.00007

SAQ 3.2d

SAQ 3.2e

Are the following statements true or false?

(*i*) The concentration should always be plotted as the abscissa of a calibration graph.

(*ii*) The expression for the fluorescence intensity is $I_f = \phi_f I_o(1 - e^{kcd})$

(*iii*) The fluorescence intensity is proportional to the concentration provided ϵcd does not exceed 0.05.

(*iv*) The limit of detection for fluorescence is usually about 10 times larger than that for absorption spectroscopy.

(*v*) In the linear region of a fluorescence calibration graph the slope is equal to $\phi_f \epsilon$ for a particular instrument.

SAQ 3.2e

3.3. THE INNER FILTER EFFECT

3.3.1. Effect on the Intensity/Concentration Relationship

We have seen that, as the concentration of a sample increases, the graph of fluorescence intensity versus concentration becomes increasingly curved because the intensity depends upon e^{-kcd} and not directly on the concentration itself. In practice, the departure from linearity is much more serious than this due to the influence of other factors. This behaviour is best illustrated by looking at some actual data. The fluorescence intensities measured for a series of solutions of quinine sulphate in aqueous sulphuric acid (0.5 mol dm^{-3}) is presented in Fig. 3.3a. The relative intensities for concentrations of up to 20 mg dm^{-3}, calculated from Eq. 3.9 and scaled so that they can be compared with the experimental values, are also shown in this table. (Above 20 mg dm^{-3} terms higher than the cubic become important.)

Π Plot both sets of data accurately on the same sheet of graph paper up to 20 mg dm^{-3} and the quinine data over the full concentration range on a separate graph. Compare the two graphs.

concentration /mg dm^{-3}	fluorescence intensity experimental	calculated
0.1	34	32
0.2	72	65
0.5	168	160
1.0	317	317
2.0	610	624
5.0	1390	1490
10.0	2210	2760
20.0	3108	4810
50.0	3406	
100.0	2750	
200.0	1105	
500.0	241	

Fig. 3.3a. *Fluorescence intensity/concentration data for quinine sulphate*

Π How does the quinine curve compare with the theoretical curve? What is the most dramatic difference between the two curves?

The two curves are very similar up to a quinine concentration of 2.0 mg dm^{-3} after which the quinine curve falls below the theoretical. The most dramatic difference is that the quinine curve passes through a maximum. Even when all the significant higher power terms are included the theoretical curve does not turn over like this so the effect is due to some other factor.

Π Over what concentration is the quinine curve essentially linear? How does this compare with value of 1.5 mg dm^{-3} we estimated at the end of Section 3.2.2?

The curve is essentially linear up to about 1.5 mg dm^{-3} which is in surprisingly good agreement with the value we obtained previously.

Π What problem would arise if you attempted to use the quinine calibration curve to determine the concentration of quinine in a completely unknown sample?

You would not know if the reading you obtained was due to a concentration above or below that corresponding to the maximum.

Π What simple practical step could you take to resolve this problem?

Dilute the sample by a factor of 2 and re-measure the fluorescence.

(The graphs obtained by plotting the data in Fig. 3.3a are shown in Fig. 3.3f. Refer to them if your answers do not tie up with the previous text.)

The chief reason for the additional curvature we observe with the quinine calibration at high concentration is that each layer of the sample absorbs some of the incident radiation so that its intensity is progressively reduced as it passes through the sample. In other words, the value of I_0, which we had previously assumed to be constant, actually decreases for most regions of the sample as the concentration increases. Let's do a simple calculation to see how much the intensity is reduced at the centre of the cell for some of the quinine solutions in Fig. 3.3a.

Π Determine the percentage of the intensity incident on the cell, I_0, which reaches the centre (ie the % transmittance of the first 0.5 cm of sample) for 1, 10 and 100 mg dm^{-3} solutions of quinine for which $\epsilon = 4540$ dm^{-3} mol^{-1} cm^{-1} (454 m^2 mol^{-1}) and $M_r = 324$.

At 1.0 mg dm^{-3}, 98.4% of the incident radiation reaches the centre of the cell. At 10 and 100 mg dm^{-3} the figures are 85.1% and 20.0% respectively.

We get these figures by first calculating the absorbance of the 0.5 cm thickness of sample through which the radiation passed to reach the centre of the cell using Beer's Law. In the case of the 1.0 mg dm^{-3} solution

$$A = \epsilon c d = 4540 \times 1.0/324 \times 0.5 = 0.0070$$

Since $A = \log(I_o/I_t)$ we have to take the antilog of the absorbance to get I_t/I_o = antilog(0.0070) = 1.016

Hence

$$I_t/I_o = 1/1.016 = 0.984 \text{ and } \%T = 98.4\%.$$

For the other two concentrations we get absorbance values of 0.0701 and 0.7010 and the conversion to transmittance takes the same course.

Thus we see that, at high concentration, the intensity of the fluorescence as predicted by Eq. 3.9 (Eq. 3.8 is not valid at these concentrations) is correct only for the first layer of sample. By the time we get to the centre of the sample it is only about 1/5 what we expected for a 100 mg dm^{-3} solution. This behaviour is one aspect of the 'inner filter effect' in which radiation is absorbed by molecules in the sample before it can excite the molecules of the fluorescent species. In the present case, the absorbing species are molecules of the analyte itself.

In practice, in fluorescence instruments using 90° geometry, the emission is collected by an optical system which is designed to exclude radiation originating near the walls of the cell – as shown in Fig. 3.3b.

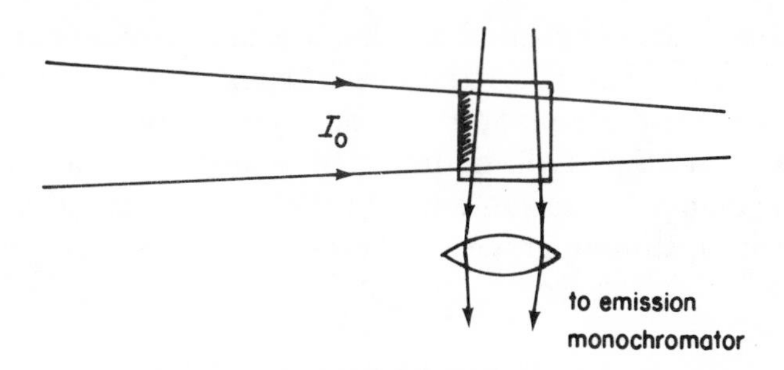

Fig. 3.3b *The effective optical paths for a 90° geometry used in fluorescence spectrometry*

Consequently, the emission monochromator does not 'see' the region of the sample shaded in Fig. 3.3b from which most of the radiation comes when we are working at high concentrations of analyte.

Π Compare the fluorescence intensity of the 100 mg dm^{-3} solution of quinine given in Fig. 3.3a with that for the 10 mg dm^{-3} solution. How far is it consistent with the reduction of I_0 to 20% at the centre of the cell as we calculated above?

The value for the 100 mg dm^{-3} solution is 1.6 times larger. If the calibration had been linear we would have expected it to be 10 times larger so the reduction is even more than our simple calculation implies. This suggests that absorption of the incident radiation is not the only factor involved – though with quinine it appears to be the major contribution. What other factors could reduce the intensity of the fluorescence?

Excited molecules may lose their energy by radiationless transfer to another molecule – in other words, quenching. With some compounds this factor is more significant and this explains the use of the terms 'concentration quenching' or 'self-quenching' to describe the fall-off in intensity as the concentration rises.

Π There may also be a loss of the *emitted* radiation by absorption before it leaves the cell. Under what circumstances would this be significant?

If the Stokes' shift were small so that the absorption band overlapped the emission band to a significant extent. This effect would be more noticeable if the emission monochromator had a wide bandwidth. This is another type of inner filter effect occurring at the emission wavelength. It is sometimes called 'self-absorption' and is analogous with a similar process occurring in flame spectroscopy.

3.3.2. Effect on the Shape of Fluorescence Bands

Apart from causing the curvature of calibration graphs in simple quantitive analysis at fixed wavelengths for excitation and emission, the inner filter effect also distorts the shape of the bands when full fluorescence spectra are run.

This is clearly evident in the emission spectrum of a 2.5×10^{-3} mol dm^{-3} solution of anthracene in propan-2-ol which is shown in Fig. 3.3c together with the absorption spectrum of a 10^{-4} mol dm^{-3} solution.

Π What differences do you notice between this emission spectrum and that of the 10^{-5} mol dm^{-3} solution of anthracene shown in Fig. 3.3d?

The 380 nm component of the emission (the 0,0 band) is missing from the spectrum of the solution of higher concentration and there is also no sign of the Rayleigh band.

Π What is the absorbance of this solution at 380 nm?

About 18 – it is 0.6 at 10^{-4} mol dm^{-3}. No wonder the 0,0 emission band can't get out!

The excitation band is also distorted by the inner filter effect, the intensity at the centre of the main excitation band being reduced relative to the wings and to the intensity of other, weaker bands. This is because the exciting radiation does not penetrate so far into the sample at wavelengths where the absorption is at a maximum with the result that these wavelengths are more seriously affected.

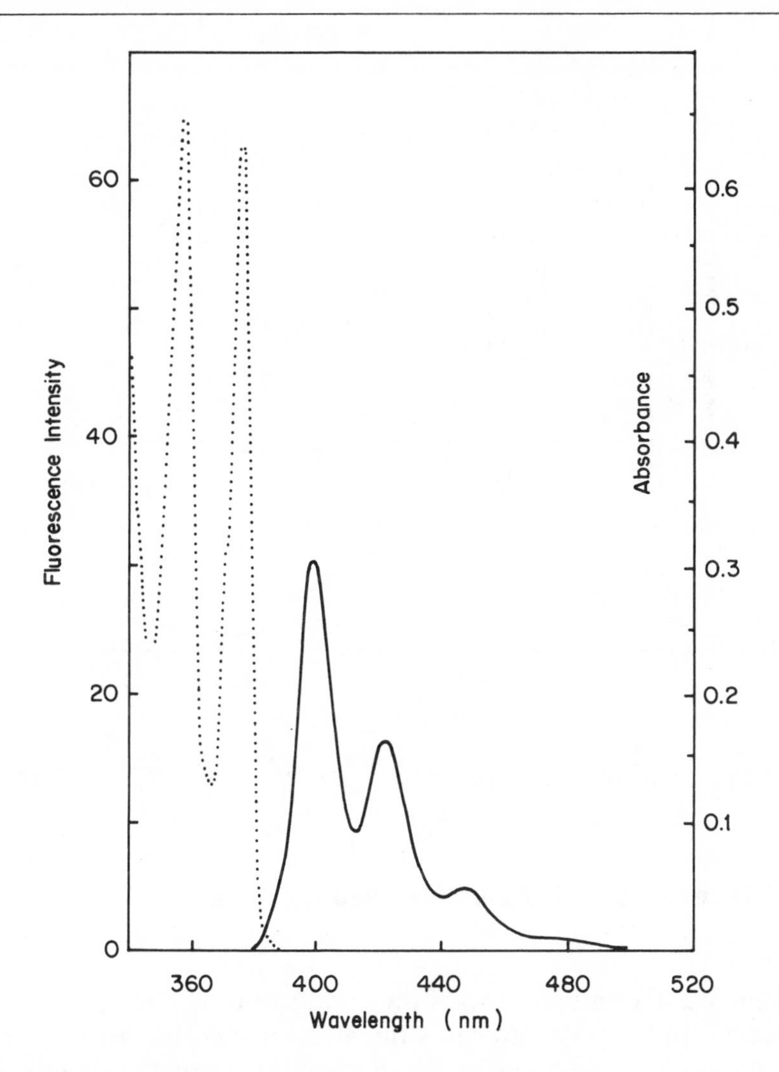

Fig. 3.3c. *Spectra of anthracene in propan-2-ol:*

Absorption spectrum (10^{-4} mol dm^{-3}) dotted line.

Fluorescence spectrum (2.5×10^{-3} mol dm^{-3}) solid line.

[Excitation at 354 nm, band-width 2.5 nm].

A particularly striking example of this is illustrated by the excitation spectrum of naphthalene in propan-2-ol recorded at a concentration of 10^{-3} mol dm^{-3} shown in Fig. 3.3e, again with the absorption spectrum of a 10^{-4} mol dm^{-3} solution superimposed.

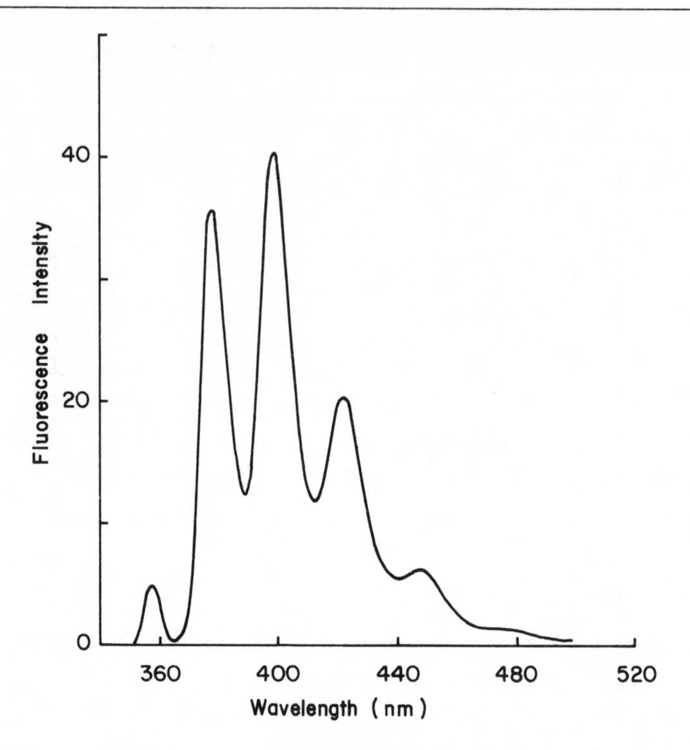

Fig. 3.3d. *Fluorescence spectrum of anthracene (10^{-5} mol dm^{-3}) in propan-2-ol*

[Excitation at 356 nm, band-width 2.5 nm]

Here the excitation band shows the contour of the absorption band in reverse. In less extreme cases the effect is merely an increase in the half-width of the band and an apparent shift in wavelength.

3.3.3. Effect from Interfering Species

The inner filter effect also occurs whenever there is a compound present in the sample with an absorption band which overlaps either the excitation or emission band of the fluorescent analyte. It becomes a problem only when the absorption is high or when the concentration of the absorbing species (and therefore its

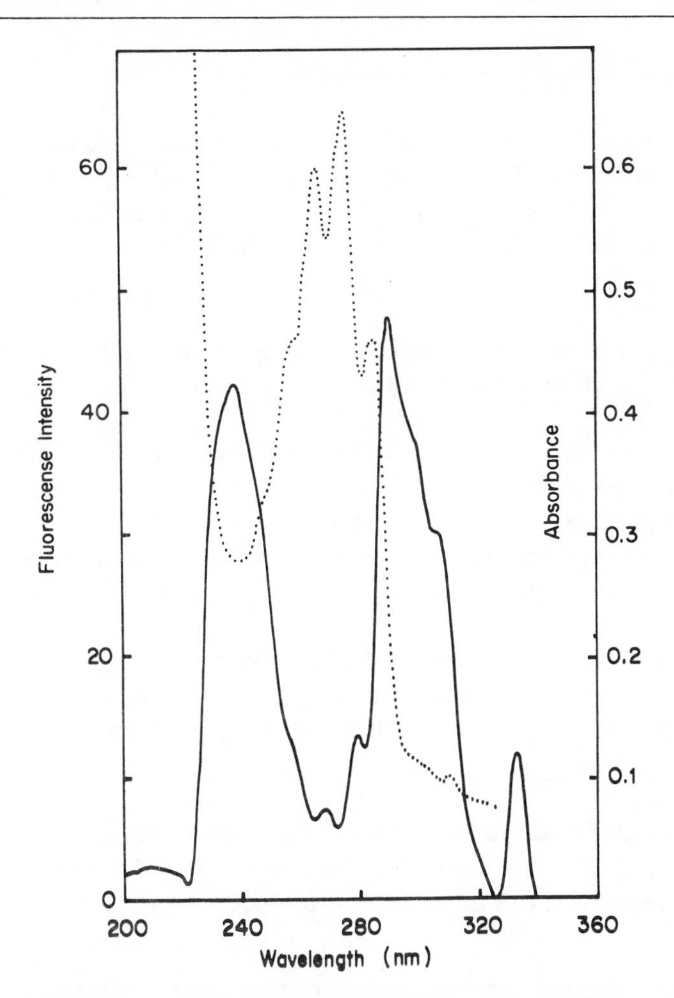

Fig. 3.3e. *Spectra of naphthalene in propan-2-ol:*
Absorption spectrum (10^{-4} mol dm^{-3}) dotted line
Excitation spectrum (10^{-3} mol dm^{-3}) solid line
[Emission at 320 nm, band-width 2.5 nm] solid line

absorbance) varies from sample to sample. The presence of a compound causing an inner filter effect will also cause some distortion of the excitation or emission spectrum of the analyte. This is often more marked than when the inner filter effect is due to the analyte itself.

3.3.4. Use of Alternative Geometries

Some of the problems associated with inner filter effect can be minimised by the use of frontal illumination or 180° (straight-through) geometry if the instrument is designed to provide these facilities. This avoids the inconvenience of having to dilute concentrated samples.

Π What is the advantage of frontal illumination and what will be its chief disadvantage?

The radiation does not have to penetrate very far into the sample. There may be problems with the incident radiation reflected from the front surface of the cell entering the emission monochromator which will therefore need to have a good stray light specification. However, the problem will not be as serious as with turbid samples.

Π The entry of exciting radiation into the emission monochromator is also a problem with 180° geometry under normal circumstances. Why will it not be a problem in the present case?

All the incident radiation is absorbed in the sample. The only problem which might arise with 180° geometry would be absorption of the fluorescence if the inner filter effect were serious at the emission wavelength.

SAQ 3.3a

What would be the effect on the fluorescence reading of diluting, by a factor of 2, the following three quinine solutions?

(*i*) 0.5 mg dm^{-3}
(*ii*) 200 mg dm^{-3}
(*iii*) 80 mg dm^{-3}

Refer to the two calibration curves for quinine sulphate in Fig. 3.3f. →

SAQ 3.3a (cont.)

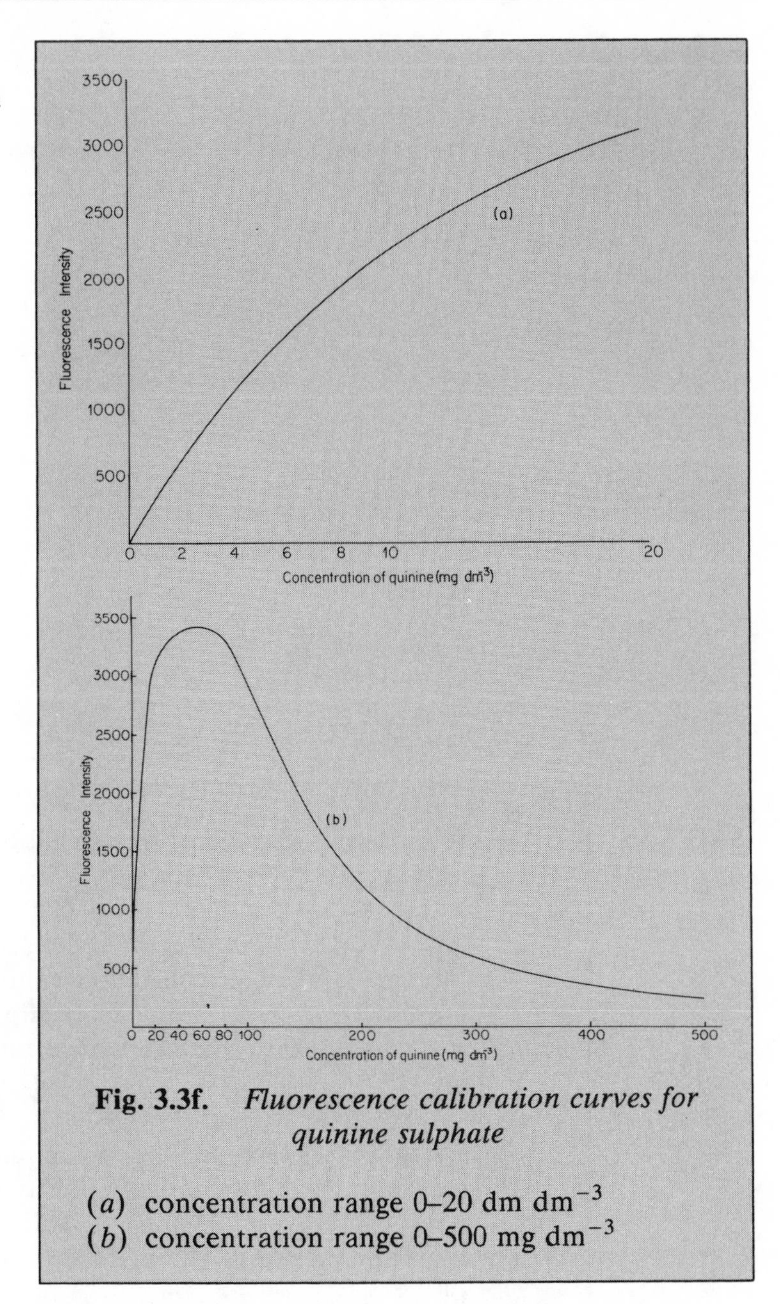

Fig. 3.3f. *Fluorescence calibration curves for quinine sulphate*

(*a*) concentration range 0–20 dm dm^{-3}
(*b*) concentration range 0–500 mg dm^{-3}

SAQ 3.3a

SAQ 3.3b

Are the following statements true or false?

(*i*) There are two contributing factors that give rise to curvature of a fluorescence intensity versus concentration calibration graph. They are the inner filter effect and the dependence of fluorescence intensity on e^{-kcd}.

(*ii*) Using 90° geometry, the maximum observed in the fluorescence intensity versus concentration calibration graph is due to the dependence of fluorescence intensity on e^{-kcd}.

⟶

SAQ 3.3b (cont.)

(*iii*) With 180° geometry the fluorescence intensity levels off at high values of the concentration without passing through a maximum provided self-absorption is negligible.

(*iv*) The profile of the excitation band is likely to be more seriously affected than the emission band by the inner filter effect originating from the analyte itself.

(*v*) An inner filter effect at the emission wavelength of the analyte due to another compound in the solution can be confirmed by measuring the absorbance of the sample at this wavelength.

(*vi*) The presence of a compound having a weak absorption at the emission wavelength of the analyte in a buffer solution used in the preparation of the sample would seriously affect the results of an analysis.

(*vii*) If the absorbance of a sample at the excitation wavelength was greater than 0.2 the sample should be diluted $\times 10$.

SAQ 3.3b

SAQ 3.3c

Examine the excitation spectra of anthracene shown in Fig. 3.3g. What *three* pieces of evidence make it clear that there is a severe inner filter effect at a concentration of 10^{-5} mol dm^{-3}? Is it still present at 10^{-6} mol dm^{-3}?

The absorption spectrum of a 10^{-5} mol dm^{-3} solution is shown in Fig. 3.3h. ⟶

SAQ 3.3c (cont.)

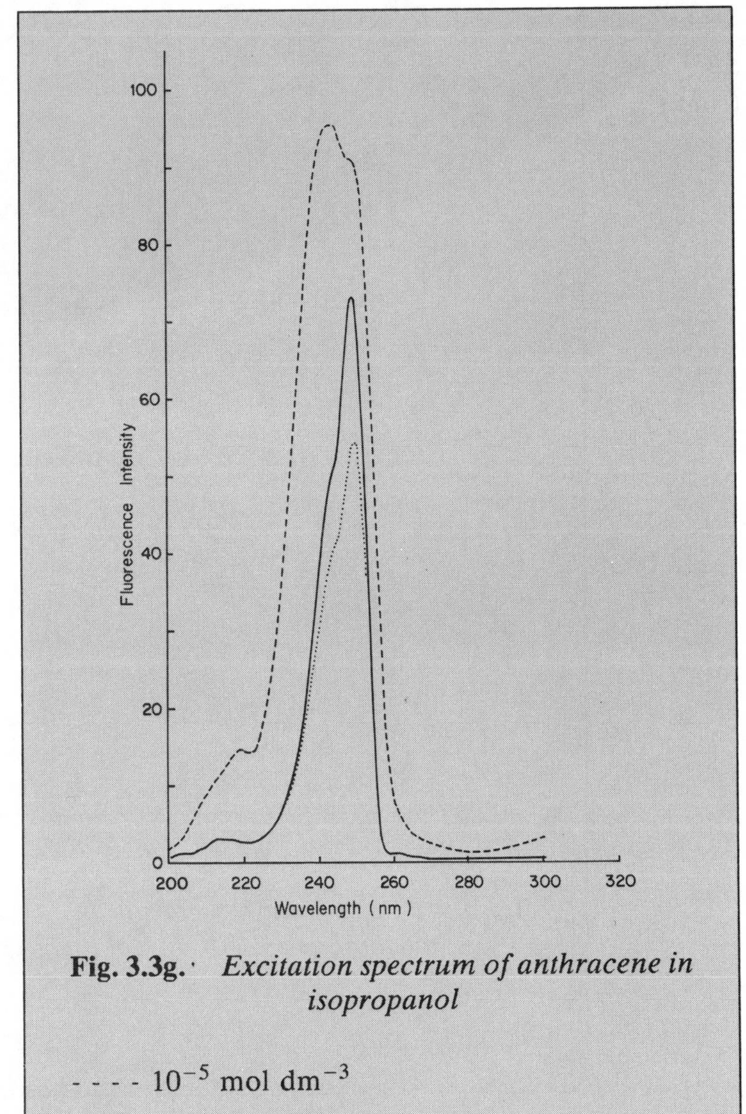

Fig. 3.3g. *Excitation spectrum of anthracene in isopropanol*

- - - - 10^{-5} mol dm^{-3}

.... 10^{-6} mol dm^{-3}

___ 10^{-7} mol dm^{-3} (gain increased × 10)

Emission at 400 nm

Bandwidth 2.5 nm

⟶

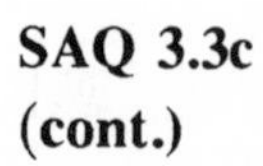

SAQ 3.3c (cont.)

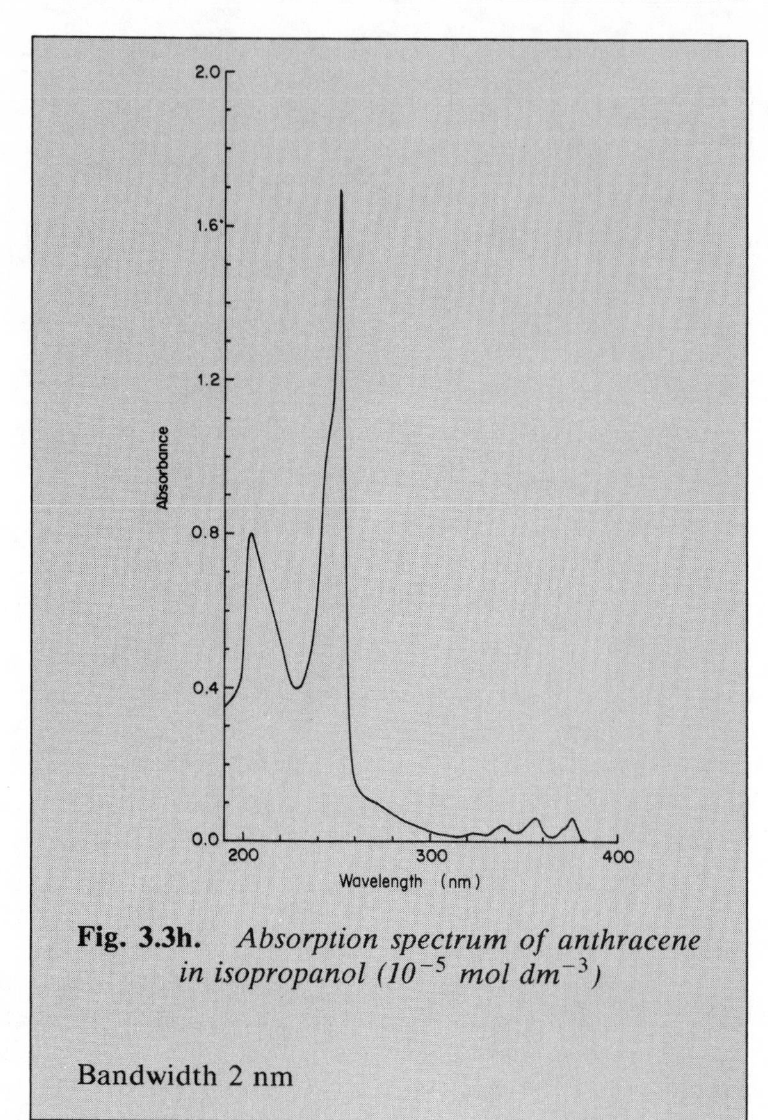

Fig. 3.3h. *Absorption spectrum of anthracene in isopropanol (10^{-5} mol dm^{-3})*

Bandwidth 2 nm

SAQ 3.3c

3.4. QUENCHING

Although the inner filter effect has the result of reducing the intensity of the radiation detected for a 90° geometry, it is not 'quenching' in the strict sense. True quenching, as we saw in Part 1 involves the removal of the energy from an excited molecule by another molecule, usually as the result of a collision. This can be important in an analysis since the fluorescence of the analyte might be quenched by the molecules of some compound present in the sample– an example of a 'matrix effect' commonly encountered in all types of analysis. If the concentration of the quenching species is constant and we know what it is, we can make allowances for it in the time-honoured way by ensuring that our standards are 'matrix-matched' to the samples. It is when the concentration of quenching species varies in an unpredictable manner that the problem becomes more acute. If the quenching effect is not too serious, one possible approach is to add an excess of the quencher to all samples and standards. This will swamp the variations present naturally and make the concentration essentially constant. Of course, this is only possible if the consequent loss of sensitivity is acceptable.

The mechanisms involved in quenching are of considerable academic interest to chemists studying the excited state of molecules. However, as analysts, we need only to note some of the principles involved so that we can be aware of the type of compound which is likely to cause trouble in a fluorescence method.

Π How do you think an excited molecule is likely to transfer its energy to a second molecule without actually emitting a photon? There are two possible mechanisms, one of which is more obvious than the other.

The more obvious method is by collision during which the energy is transferred directly. We might represent this by the equation

$$A^* + Q \rightarrow A + Q^*$$

where A is the fluorescent analyte molecule and Q the quenching species. The other process involves the formation of a complex between the analyte and quenching species which can then degrade the energy by some internal conversion mechanism

$$A^* + Q \rightleftharpoons AQ^* \rightsquigarrow AQ$$

Π Why are these two processes dependent on the concentration of the quenching species, Q?

In the case of the collision mechanism, the higher the concentration of Q the greater the chance of a collision occuring between the analyte and the quencher and so the greater the chance of de-exciting A^*.

With complex formation we have a simple mass action effect which shifts the equilibrium further to the right (formation of AQ^*) as the concentration of one of the reacting species is increased.

In general it is difficult to predict whether the fluorescence of one compound is likely to be quenched by the presence of another specific known compound – unless we know that the two molecules react to form non-fluorescent species. However, one general principle that can be applied is that molecules which can induce the excited fluorescent species to pass over into the triplet state, usually as a result of formation of a loose complex, will act as quenching agents. By analogy with the principles discussed in Part 1 affecting the fluorescence efficiency of organic compounds, we shall therefore expect compounds containing heavy atoms (especially the heavier halogens) to be efficient quenching agents.

Compounds containing unpaired electrons can also act as efficient quenching agents. The most important compound of this type is molecular oxygen which is well known as a paramagnetic species and is, of course, universally present in solution in all samples. (The solubility in most solvents at room temperature is about 10^{-3} mol dm^{-3} – several orders of magnitude higher than the analyte in most fluorescence determinations). It is particularly effective in removing the energy from triplet state molecules and is therefore a particular problem in phosphorescence and with molecules with relatively long fluorescent lifetimes. In other cases its effect is only a few per cent but still enough to be significant. Consequently when developing any new fluorimetric analytical method it will be necessary to check whether dissolved oxygen affects the fluorescence intensity. This is most readily done by removing the oxygen either by bubbling oxygen-free nitrogen through the sample or by heating it, possibly under reduced pressure. If either of these two processes causes the fluorescence intensity to increase it will be necessary to include them in the analytical procedure.

SAQ 3.4a

State whether each of the following statements is true or false:

(*i*) Fluorescence is quenched when a molecule in an excited electronic state is deactivated without the emission of a photon.

(*ii*) The fluorescence efficiency of a compound in solution is affected by the presence of other compounds.

(*iii*) Quenching can be used to measure the concentration of the quenching species.

⟶

SAQ 3.4a (cont.)

(*iv*) Self-quenching results in the intensity of fluorescence decreasing as the concentration increases.

SAQ 3.4b

How could you demonstrate that the reduction of the intensity of the fluorescence of quinine sulphate at 440 nm by sodium chloride is due to quenching and not to the inner filter effect?

SAQ 3.4c

Fill in the blanks in the following paragraph using the words listed below. Each word may be used once only.

Both quenching and the inner filter effect result in the .... of the intensity of fluorescence. They differ in that quenching does not involve .... of radiation while the inner filter effect does not involve .... state molecules. With quenching, excited state molecules transfer their .... to another molecule by .... or following the formation of a .... or by some other means of .... transfer. Other species present in the sample can give rise to the inner filter effect only if they .... radiation at the .... of excitation or .... The effect of quenching on an analysis can be corrected by adding an .... of the species responsible to all samples unless the .... of the analyte is near the .... limit.

energy	wavelength	complex
absorb	excited	detection
excess	reduction	emission
collision	absorption	concentration
radiationless		

3.5. PHOTODECOMPOSITION

Although we have previously regarded photodecomposition as an alternative fate of molecules in the excited state and a process which competes with the emission of fluorescence, it can have a significant effect in an analysis and this is an appropriate time for a few words on the subject.

Photodecomposition can result in either an increase or a decrease in the observed intensity from a sample.

Π Try to think of processes which can result in each type of behaviour.

A decrease in the intensity would be observed if the fluorescent species decomposed into non-fluorescent products. An increase in the intensity would be observed if the fluorescent species decomposed to give products which were more fluorescent than the original species under the same conditions of excitation and observation. Another possibility is that some other component present in the sample such as the fluorimetric reagent (which is always in excess over the analyte) may decompose to give fluorescent products. Other effects may arise from the destruction or formation of quenching agents in the sample through photodecomposition.

Π How could you tell that photodecomposition was occurring in your sample?

The most usual indication of photodecomposition is that the fluorescence intensity varies with time – up or down as we have seen. Confirmation can generally be obtained by comparing the absorption spectrum before and after irradiation. If the fluorescent species or the decomposition products are coloured you may even be able to detect a visible change in the sample. If you are using a spectrofluorimeter you could of course run the fluorescence spectra which would change as you were recording them.

Π How might you minimise the extent of photodecomposition?

The most obvious method is to limit the duration of the exposure of the sample to the exciting radiation. Fluorescence instruments are always provided with a shutter on the excitation side of the optical system and this should be kept closed until you are ready to take the measurement which should be taken as quickly as possible. It also helps to use radiation with the longest possible wavelength and with the lowest possible intensity to excite the fluorescence even though this may mean operating with reduced sensitivity.

Another less obvious remedy which is often effective is to remove dissolved oxygen. The oxidation of many organic molecules with molecular oxygen is frequently enhanced with uv radiation so removal of oxygen may well be beneficial in this respect if it has not already been employed to avoid its quenching action. The fluorescent complex formed between borate ions and benzoin is a very good example of the problems caused by dissolved oxygen on both counts.

Apart from these measures there is very little you can do to combat photodecomposition. Many organic compounds decompose under exposure to uv radiation because the photon energy is comparable with the bond dissociation energy of many chemical bonds. Large molecules and conjugated molecules, of which dyestuffs are examples, are particularly susceptible to photodecomposition and we are all well aware of the problems associated with the fading of many dyes when exposed even to visible radiation. In the long run, the fluorescent species used used in fluorimetry and any reagents used to form them have just got to be stable to radiation of the wavelength required to excite the fluorescence.

SAQ 3.5a

Select the correct ending to the following sentence:

Radiation of wavelength 400 nm is less likely to cause photodecomposition of an organic molecule than radiation of wavelength 300 nm because

(*i*) its photon energy is higher.
(*ii*) it is visible radiation.
(*iii*) its photon energy is lower.
(*iv*) it is more readily absorbed.

SAQ 3.5a

SAQ 3.5b

Are the following statements true or false?

(*i*) There is no need to outgas a solution for fluorescence measurement because oxygen quenching occurs with the triplet state.

(*ii*) If the fluorescence of a solution decreases whilst it is being measured it is likely that photodecomposition is taking place

(*iii*) Fluorescence and photodecomposition are competitive processes because both result in the deactivation of excited state molecules.

(*iv*) Photodecomposition of the analyte always reduces the intensity of fluorescence.

SAQ 3.5c

Which of the following do *not* provide a possible cause of photodecomposition?

(*i*) The photon energy of uv radiation is comparable with the bond dissociation energy (per molecule) in many organic molecules.

(*ii*) Many compounds react with oxygen under strong uv irradiation.

(*iii*) Absorption of uv radiation can take a molecule into a high electronic excited state.

(*iv*) An excited molecule can lose its energy by collision with another molecule.

SAQ 3.5d

Which of the following steps would *reduce* the extent of photodecomposition?

(*i*) Increase the slit-width of the excitation monochromator.

(*ii*) Keep the excitation shutter closed until you are ready to take the reading.

(*iii*) Bubble oxygen-free nitrogen through the sample.

(*iv*) Reduce the slit-width of the emission monochromator.

(*v*) Use a shorter wavelength for excitation.

SAQ 3.5e

The bond dissociation energy of the C—Br bond is 290 kJ mol^{-1}. The uv absorption spectrum of tribromomethane is shown in Fig. 3.5a. Determine whether the photon energy of radiation of wavelength

(*i*) 450 nm
(*ii*) 350 nm and
(*iii*) 250 nm

is sufficient to decompose the molecule and state whether decomposition will occur at each wavelength.

1 mole contains 6.0×10^{23} molecules
$h = 6.6 \times 10^{-34}$ J s
$c = 3.0 \times 10^{8}$ m s^{-1}

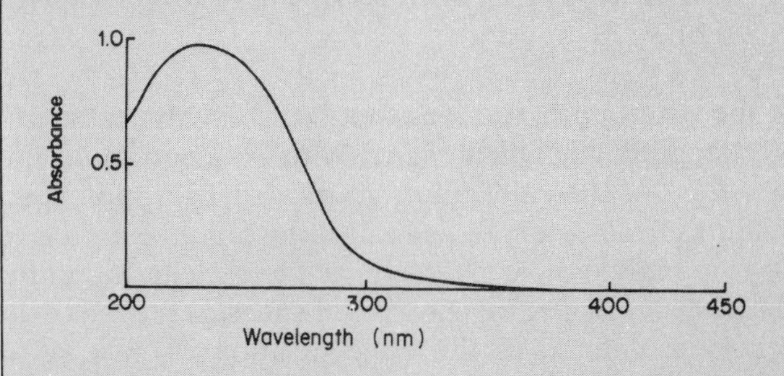

Fig. 3.5a. *Ultraviolet absorption spectrum of tribromomethane*

SAQ 3.5e

3.6. SENSITIVITY, LIMIT OF DETECTION AND THE BLANK

In the strict sense, the sensitivity of an analytical method is the rate of change of analytical signal with concentration, in other words, the slope of the calibration graph. In absorption spectroscopy this definition can be expressed as the absorbance for a given concentration of analyte or, conversely, as the concentration to give a stated absorbance. In atomic absorption spectroscopy for example the sensitivity is defined as the concentration to give an absorbance of 0.0044 (ie 1% absorption or 99% transmittance).

Π It is not possible to define sensitivity for an emission technique in similar terms. Why do you think this might be?

The absorbance of a sample is determined by the properties of the sample and is independent of the instrument with which the measurement is made. In emission, the analytical signal is an intensity value which depends not only on the sample but also on a number of instrumental factors such as source intensity and optical efficiency.

Π Assuming linear expressions of the form $y = mx$, what is the slope of the calibration graph in

(*a*) absorption spectroscopy when plotting absorbance versus concentration, and in

(*b*) fluorescence spectroscopy when plotting fluorescence intensity *vs* concentration in the linear region?

(*a*) ϵd If Beer's Law holds, $A = \epsilon c d$, $y = A$, $x = c$ and $k = 0$.

(*b*) $2.303\ I_0 \phi_f \epsilon d$ from Eq. 3.8.

Π Identify the quantities in this answer which relate to

(*a*) the actual analyte or the sample used, and

(*b*) the instrument

(*a*) In absorption spectroscopy, the absorptivity, ϵ, is a property of the sample whilst d is determined by the size of the cell chosen for the analysis. In fluorescence, the gradient is also determined by ϵ and d but, in addition, the fluorescence efficiency of the compound, ϕ_f, will play a major part.

(*b*) There is no instrumental factor involved in the gradient of the absorption calibration graph. In fluorescence, the incident intensity, I_0 is also important. The expression for I_f actually gives the total fluorescence of the sample, and for a particular instrument it will also be necessary to include factors to take account of the proportion of the fluorescence collected by the detector and also the sensitivity of the detector at the wavelength of the emission.

3.6.1. Detection Limits

The term sensitivity is often used in a rather vague sense as a measure of the performance of a particular technique at very low concentration of analyte. An alternative quantity which can be precisely defined for both emission and absorption techniques is the *detection limit*. This is defined as the concentration of analyte which gives rise to an analytical signal equal to twice the noise. Alternatively the detection limit may be defined as the concentration equivalent to twice the standard deviation of at least ten readings on an analyte sample at a concentration just above the blank level. This definition is in fact exactly the same as the signal/noise definition but it is perhaps a more realistic approach to the detection limit from the analytical point of view and can be applied to any analytical technique including the classical methods.

Π There is a fundamental difference in the way the analytical signal is measured in emission and absorption spectroscopy which becomes important as the concentration of the sample approaches the detection limit. What is this? Would you expect this to make emission techniques *more* or *less* sensitive than absorption techniques?

The emission technique is considerably *more* sensitive because it is intrinsically easier to measure a small signal against zero background accurately than to compare two large signals.

The limit of detection of the fluorescence technique is generally of the order of 1000 times lower than that for uv absorption spectrometry.

Of course, the limit of detection is not the only factor which determines the sensitivity of an analytical technique. We shall see shortly that background emission from the blank is a serious problem in fluorescence work and often determines the lower limit of concentration that can be reached. Another problem is that the precision of measurement of the analytical signal falls off rapidly as we approach the detection limit due to instrument noise and other factors. This was the reason why, in Section 3.2.3, it was suggested that the 'lowest measureable concentration' should be taken as ten times the detection limit for all practical purposes.

3.6.2. Factors Affecting the Sensitivity of Fluorescence

In discussing the sensitivity of fluorescence procedures in analysis it is convenient to separate the contributions from the properties of the fluorescent molecule itself, the performance of the instrument and the chemistry involved in the preparation of the sample. These are referred to as the *absolute sensitivity*, the *instrumental sensitivity* and the *method sensitivity* respectively. Let's have a look at each of these types of sensitivity in turn.

The absolute sensitivity is determined chiefly by the molar absorptivity and the fluorescence efficiency of the analyte molecule itself. A high fluorescence efficiency is usually associated with some rigidity in the molecular structure which is often brought about by the presence of aromatic rings and other conjugated systems. These factors also result in a high value of the absorptivity so that, if a molecule has a high fluorescence efficiency, its absorptivity is also likely to be high. The converse is not necessarily true because other features in the electronic structure of the molecule such as the presence of lone pairs or heavy atoms seriously reduce the fluorescence efficiency but do not affect the absorptivity.

Π The instrumental sensitivity will determine the detection limit for a particular compound on a particular instrument. We discussed some of the factors involved, in Part 2. See if you can now draw up a list of those which will contribute to the instrumental sensitivity of a fluorescence spectrometer.

Your list should include the following:

1. Source intensity.

2. Efficiency of the optical system used to irradiate the sample.

3. Efficiency of the optical system used to collect the radiation emitted by the sample.

4. Aperture of the monochromators.

5. Band-pass of the monochromators.

6. Sensitivity of the detector.

7. Noise level in the detector circuit – a low value will allow for a much higher degree of amplification.

Π Which of these factors is wavelength-dependent?

Source intensity and detector sensitivity are the obvious ones. The energy throughput of a grating monochromator also varies with wavelength being at a maximum at the blaze wavelength and falling off quite rapidly on either side of it. The transmittance and reflectance of other optical components are also dependent on the wavelength, particularly if they have been surface-coated to reduce light losses. Filters, of course, only operate at a single wavelength.

The effect of the monochromator band-pass requires a little more consideration.

Π How would you increase the band-pass of a grating monochromator?

By increasing the slit-width. Most instruments provide a choice of values. The band-pass of a filter instrument is not adjustable – unless, of course, a different filter is available.

(It will avoid confusion if we use 'band-pass' in this context and reserve 'band-width' for describing bands in spectra.)

Increasing the slit-width of the excitation monochromator lets more light through but to an extent which depends on the circumstances. With a continuous source like the xenon arc where radiation is present across the entire spectral band-pass the intensity increases as the square of the band-pass. With a line source like the mercury lamp the increase is only linear because the line width is very much less than the band-pass.

A similar situation arises with the emission monochromator and this provides a useful way of discriminating between lines or narrow bands and broad bands as we shall see later. With both monochromators the most efficient procedure is to operate with the entrance and exit slits of equal width – indeed it is not usually possible to vary them independently.

In practice, instrument manufacturers commonly express the sensitivity of their instruments by quoting the detection limit for a particular fluorescent compound – usually quinine sulphate. The figure is usually given in 'ppb', ie parts per billion – the American version meaning 1000 million. A typical value is 0.005 ppb which means that the signal from a solution containing 0.005 μg dm^{-3} should be twice the noise level. This value will depend upon the band-pass at which the instrument is operated. Unfortunately this may differ between different instruments and is sometimes not quoted at all which makes comparison difficult.

A more precise standard has recently been adopted by some manufacturers. This is to quote the signal/noise ratio for the Raman band of water observed at maximum sensitivity settings with a band-pass of 10 nm. The Raman effect is discussed briefly in the following section. It is important in fluorescence because it interferes with the observation of weak fluorescence bands. For the present purpose however we need only note that when a sample of water is irradiated at 350 nm its Raman band appears at 395 nm. These are the wavelengths used to produce the signal to which the noise level is compared. For a typical modern dual monochromator spectrofluorimeter this signal/noise figure is greater than 50.

The method sensitivity takes account of preconcentration steps in the preparation of the sample on the one hand and the limitations imposed by the fluorescence of the blank on the other. The former is a routine analytical procedure which you have probably already met. A typical example of this is the extraction of an analyte from aqueous solution into a smaller volume of an organic solvent such as cyclohexane. The effect of the blank is a critical factor in fluorescence work and merits a more detailed discussion in a separate section.

3.6.3. The Blank

Although measurements on a pure solution of a fluorescent compound may indicate that it should be possible to measure its concentration down to a very low level determined by the noise, in practice it is usually found that such estimates are over-optimistic. The common experience is that, as the gain is increased to amplify the signal, the background to the signal rises so that it becomes impossible to keep the reading on scale, even when the zero of the instrument is off-set. Most commercial instruments are provided with a 'zero-suppression' facility which allows the zero to be shifted to a considerable extent – twice the full-scale deflection of the read-out device or more. Fluorescence measurements at low concentrations are therefore more often found to be 'blank-limited' rather than 'noise-limited'. Since this limitation arises from the nature of the sample or from the treatment it receives prior to measurement, you can understand why the term 'method sensitivity' is used to describe the practical sensitivity likely to be achieved in a real situation.

There are many factors present in a typical sample which contribute to the blank. Some of these are associated with impurities while others are due to phenomena originating from the solvent. Since solvent molecules outnumber solute molecules by a huge factor (often more than 10^8) weak signals arising from secondary effects in the solvent will eventually swamp the fluorescence from the solute.

Π How do you think that impurities can give rise to high blank values? (There are two distinct processes.)

The most serious effect comes from impurities which are themselves fluorescent and emit radiation which has a similar wavelength to that emitted by the analyte. The problem is made worse if both the excitation and emission wavelengths are similar which makes discrimination by careful selection of wavelengths difficult, even with a dual monochromator instrument. The second type of impurity is one which is present as suspended particles which scatter radiation at the excitation wavelength.

Π Why should scattered radiation give rise to a high blank value? Surely the emission monochromator is set at a longer wavelength and should reject it.

If the intensity of the scattered radiation is very high compared with that of the fluorescence it will give rise to a high stray light level which will swamp the analytical signal.

Some fluorescent impurities will be present as a matter of course in samples acquired from natural sources. Others, however, may be introduced during the preparation of the sample for analysis. Several steps may be involved including adjustment of pH, dilution and addition of a fluorimetric reagent. As we have seen, in some cases it may be necessary to extract the analyte species into an organic solvent.

Π In each of these steps there is a possibility of introducing fluorescent impurities. Try to draw up a list of the possible source of these impurities and suggest ways of avoiding them.

Your list might include some or all of the following:

(*a*) Acids, bases and buffer reagents used to adjust the pH. These will have been prepared by commercial processes which provide ample opportunities for contamination by trace impurities. For fluorescence work they should be carefully purified by distillation where practical. Recrystallisation is less efficient because impurities tend to be adsorbed on to the surfaces of the crystals. A more successful treatment is to boil the reagent solution with activated charcoal to adsorb the impurities before filtering and recrystallising.

(*b*) Water used as diluent or solvent for reagents. This needs to be specially purified by triple distillation in clean glass or copper stills before use. Deionised water prepared using an ion exchange process is less satisfactory because it may contain organic materials picked up from the resin. It should not be stored in plastic containers because of the danger of leaching out fluorescent species.

(*c*) The fluorimetric reagent itself. Commercial organic compounds particularly the complex dye-stuffs often used as fluorimetric reagents, are never pure and may require extensive purification by special techniques (such as liquid chromatography) if the impurities are fluorescent. Trace metals may also be troublesome, particularly the metal for which the reagent is to used and others forming fluorescent complexes with it. The common elements like calcium and magnesium are a particular problem because of ease with which they are picked up from the environment. Ion exchange provides a possible method for removing them.

(*d*) Organic solvents. Although these can be obtained in a high state of purity (special 'spectroscopic grade' materials are produced commercially), they are likely to become contaminated during transfer because of their solvent properties for grease which is easily picked up from dirty glassware or the fingers of careless laboratory staff. This can lead to a high background emission from these solvents. Careful redistillation or purification by liquid chromatography is a possible remedy though it is probably cheaper in the long run to throw away suspect material and open a fresh bottle.

Π What about the problem of suspended particles? What might these be and where might they have come from?

Most natural samples contain particles of many types and sizes. River samples may contain silt and particles of decaying vegetation, sea water may also contain micro-organisms (which may also be bioluminescent to add to our problems!) and biological fluids may contain fragments of cells and tissue as well as large molecules such as proteins which are frequently present in colloidal form.

Solutions of reagents made up under laboratory conditions are less likely to give trouble unless they have been allowed to stand open to the atmosphere when they may have picked up particles of dust, hairs etc. In some cases a reagent may decompose slowly on standing and produce insoluble products.

Perhaps you may have thought of other examples of suspended

solids. This list is by no means exclusive! How would you set about removing them?

The larger size particles, dust and hairs can be removed by filtration using filter papers or sinters. Smaller particles can be removed using ultra-fine filtering media such as membrane filters. There is always the danger that traces of fluorescent impurities may be picked up from the filter so care must be taken in the selection of suitable materials. Alternatively, the particles may be removed using a centrifuge though this is likely to be a more lengthy and less efficient procedure.

Whatever treatment is used there will usually be a residue of very fine or colloidal particles remaining in the sample. If this proves troublesome it may be possible to remove it by adsorption on to active charcoal or other surface-active media.

Π Although this method is very effective for cleaning up reagents which are present at relatively high concentrations, there is a serious risk in applying it to the actual sample.What do you think this might be?

Since the analyte is present at exceedingly low concentration it might very easily be removed along with the impurities.

As you can see, obtaining satisfactory results from very dilute solutions in the presence of suspended particles can involve the operator in time-consuming and tedious (and costly) bench work. Such sample pre-treatment is always to be avoided whenever possible and it is a great advantage if the instrument used for the fluorescence measurement has a very high stray light specification so that it can cope with dirty samples. The higher initial cost of the instrument is quickly recovered in a laboratory handling large numbers of such samples.

It is not only solid particles which give rise to scattering of the incident light. Bubbles of air in the sample behave similarly and are a particular problem as they pass in and out of the light path as they are formed and released.

The problem can be avoided by degassing the sample by heating it or by applying a vacuum to it and by making up reagent solutions with de-aerated water.

You may be getting the impression that fluorimetric analysis at very low trace levels is fraught with problems and requires very great care on the part of the operator, so why bother? You should remember however that fluorimetry becomes the method of choice for many analyses for the very reason that it is so sensitive and can be used at levels well below those accessible by other techniques – it is the low level of analyte that causes the problems, not the technique. For routine analysis at the very low levels to which fluorescence is sensitive, a very tight laboratory regime is maintained to avoid contamination. As we have seen, great care must be taken in preparing the sample for analysis and in the selection preparation and storage of reagents. Apart from this however similar precautions must be taken in maintaining apparatus with which the sample comes into contact.

Π It goes without saying that all glassware must be scrupulously clean and stored in a clean dust-free environment. However, some thought must be given to the method of cleaning which may be quite adequate for normal laboratory procedures but may give rise to unforeseen problems in low level fluorescence work. What might these problems be?

The use of organic reagents and certain detergents may leave fluorescent deposits on glassware which eventually gets into samples and so it is necessary to be particularly careful to rinse apparatus several times finishing with ultra pure water. The use of a glass cloth to dry apparatus also constitutes a hazard since the cloth may well contain materials such as optical brighteners introduced during manufacture or later laundering which again could eventually find their way into samples, and give a high blank value. Some grades of paper tissue also suffer from this problem.

Yet another source of interference comes from the cuvette itself.

Π Synthetic quartz has a very low blank level and is normally entirely satisfactory when clean. How can careless handling of cells cause problems?

Traces of fluorescent impurities may be left on the optical surfaces. The fluorescence emission produced by these may be scattered and enter the emission monochromator – particularly if the sample contains suspended particles giving a high scattered background. Of course, on occasions when it is necessary to use frontal illumination it is essential that the cell be absolutely clean.

Cheap plastic cells are likely to give a rather higher fluorescent background than silica or glass if used near their wavelength limit (320 nm). They are intended for use at longer wavelengths and for routine analysis where sensitivity is not a problem. In this type of application they are quite satisfactory.

Even if we achieve absolute perfection in the preparation of a sample which is crystal clear and completely free of any fluorescent impurities we still have the problem of the contribution from the solvent molecules themselves to contend with.

Π There are two types of scattering at the molecular level which were mentioned in Section 2.3. Can you remember what they are? What are the differences between these two types of scattering? Which is the stronger? How do they contribute to the blank?

The two processes referred to are Rayleigh and Raman scattering. Rayleigh scattering occurs at the same wavelength as the exciting radiation and is much more intense than Raman scattering which occurs at a slightly longer wavelength. Like scattering from particles, Rayleigh scattering adds to the stray light level of the instrument since the monochromator is set to the fluorescence emission wavelength so that the Rayleigh scattering does not reach the detector directly.

Raman scattering, though weaker than Rayleigh scattering, is at longer wavelength and so behaves more like the emission from a fluorescent impurity which overlaps the emission band. It will therefore be more difficult to discriminate between the fluorescence emission of the analyte and the Raman emission of the solvent. Hence the latter is likely to impose the ultimate limitation on the lowest concentration of the analyte which can be observed (unless the sensitivity of a particular instrument is limited).

Raman spectra arise from a scattering process in which energy is transferred from the incident photons to the scattering molecules so that the molecule ends up in an excited vibrational state. Raman spectra are therefore vibrational spectra observed in the visible and uv regions of the electromagnetic spectrum to the long wavelength side of the (monochromatic) radiation which excites them. The 'Raman shifts' of the bands in the Raman spectrum, $\overline{\nu}_0 - \overline{\nu}_r$, where $\overline{\nu}_0$ is the wavenumber of the exciting radiation ($1/\lambda_{ex}$ in cm^{-1}) and $\overline{\nu}_r$ are the wavenumbers of the Raman bands, frequently correspond to the wavenumbers of bands in the infrared spectrum.

Π Since Raman spectra are associated with a particular solvent and are observed as a series of bands whose displacement from the exciting radiation is constant and are of narrow bandwidth there are three possible ways of reducing their interference with fluorescence. Can you deduce what they are?

(*a*) Change the solvent to one whose Raman bands do not coincide with the fluorescence emission maximum.

(*b*) Change the excitation wavelength to shift the Raman bands away from the fluorescence band.

(*c*) Increase the slit width. The fluorescence band intensity will increase as the square of the band-width. The narrower Raman band will increase in intensity only linearly with the band-width which will favour the observation of the fluorescence band – though it may create other problems with a dirty sample.

Π What effect will changing the excitation wavelength have on the wavelength and intensity of the fluorescence emission band?

There will be no effect on the emission wavelength. There will be a drop in the intensity however since we shall have moved away from the excitation maximum and this will reduce the sensitivity and impose a limit on how far we can shift the excitation wavelength. Increasing the slit width will help to restore the intensity but may result in problems if other types of scattering or interference are present.

SAQ 3.6a

Are the following statements true or false?

(*i*) The absolute sensitivity of a particular compound is given by the slope of the calibration graph.

(*ii*) The instrumental sensitivity stated as '0.005 ng dm^{-3} of quinine' means that the instrument is capable of recording that concentration of quinine with a signal/noise ratio of 50.

(*iii*) The instrumental sensitivity is now commonly expressed in terms of the signal/noise ratio of the Raman band of water excited at 350 nm with a band-pass of 10 nm.

(*iv*) The method sensitivity is generally governed by the emission from the blank.

SAQ 3.6b

Identify the item which is out of place (in the context of fluorescence sensitivity) in each of the following groups:

(*i*) Incident radiation intensity, fluorescence efficiency, detector sensitivity, monochromator band-pass.

(*ii*) Incident intensity, detector sensitivity, monochromator transmittance, monochromator aperture.

(*iii*) Raman scattering, Rayleigh scattering, stray light, Tyndall scattering.

(*iv*) Buffer solution, fluorescence reagent, organic solvent, colloidal solution.

(*v*) Distillation, membrane filtration, liquid chromatography, charcoal treatment.

(*vi*) Change wavelength of excitation, use frontal illumination, increase band-pass of emission monochromator, change solvent.

SAQ 3.6c

Rearrange List B so that each item becomes a remedy for the corresponding problem encountered in sample preparation given in List A.

List A	*List B*
Metal impurities in fluorescence reagents	Redistillation
Bubbles of gas	Filtration
Fluorescent impurities in solvents	Charcoal treatment
Fluorescent impurities in a buffer solution	Heat the sample
Suspended particles	Ion exchange

SAQ 3.6d

Rearrange List B so that each item becomes a remedy for the corresponding sampling problem in list A.

List A	*List B*
High blank	Increase emission slit-width
Raman interference	Use auxiliary emission cut off filter
High stray light	Use frontal illumination
Strong inner filter effect.	Use zero suppression

SAQ 3.6e

The main infrared absorption band of water occurs at 3300 cm^{-1}. Calculate the wavelength of the Raman band when an aqueous sample is exposed to exciting radiation at 390 nm during a fluorescence experiment.

SAQ 3.6f

Emission spectra of a solution of anthracene in propan-2-ol (10^{-7} mol dm^{-3}) recorded with the emission slit width set to 2.5, 5, 10 and 20 nm are shown in Fig. 3.6a. The excitation slit-width was 2.5 nm throughout and the excitation wavelength was 354 nm. Raman spectra of propan-2-ol recorded under the same conditions are shown in Fig. 3.6b. The ordinate expansion factor was adjusted for each spectrum to maintain the peak intensities approximately constant. This is recorded on the spectra. ⟶

SAQ 3.6f (cont.)

(*i*) How does the Raman spectrum interfere with the fluorescence band?

(*ii*) Measure the peak height of the Raman band and the strongest fluorescence peak in each spectrum. Divide all these readings by the ordinate expansion factor (f), printed on each spectrum, to bring them to the same intensity scale. Does increasing the slit-width improve discrimination in favour of the fluorescence peak?.

(*iii*) Refer to the excitation spectrum of anthracene, Fig. 3.6c. Suggest an alternative excitation wavelength to remove the Raman interference. How much loss in intensity would result from this change?

(*iv*) Check whether the use of carbon tetrachloride with Raman bands at shifts of 459 and 776 cm^{-1} would be a more effective way of removing the interference.

(*v*) Which ir band of propan-2-ol does the strongest Raman band correspond to?

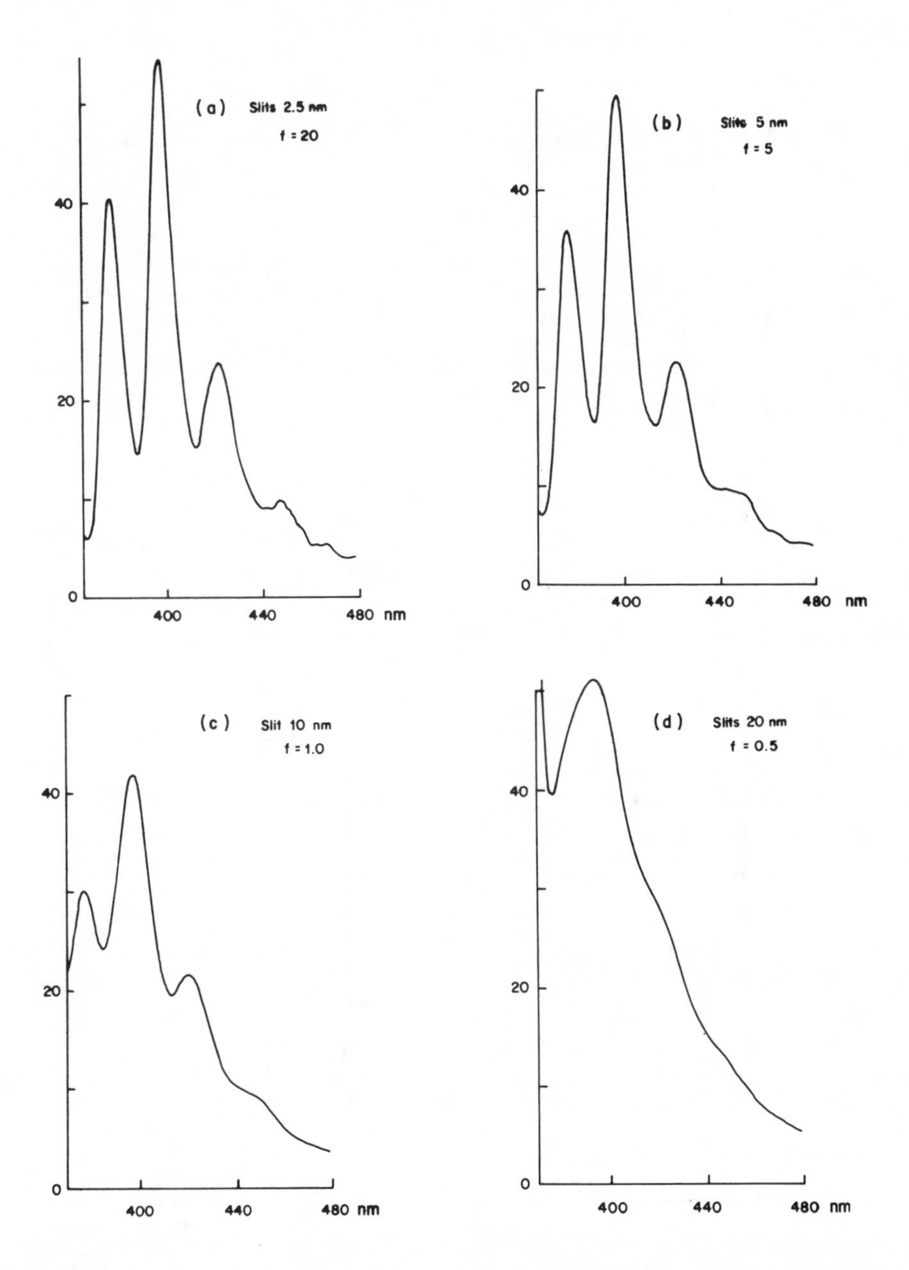

Fig. 3.6a. *Fluorescence emission spectra of anthracene in propan-2-ol (10^{-7} mol dm^{-3}), with excitation at 354 nm and an excitation slit width of 2.5 nm. [Emission slit width and ordinate expansion factors, f, are given on the spectra.]*

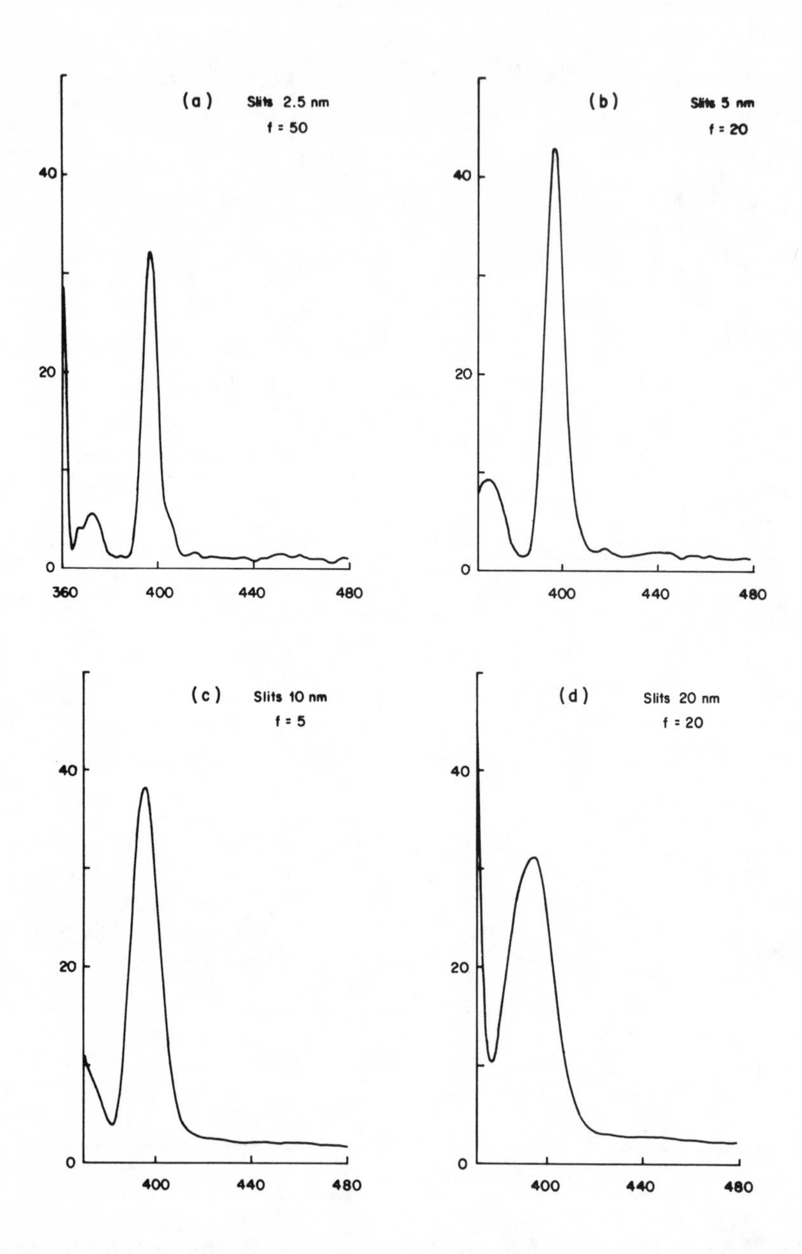

Fig. 3.6b. *Raman spectra of propan-2-ol, with excitation at 354 nm. [Emission slit width and ordinate expansion factors, f, are given on the spectra.]*

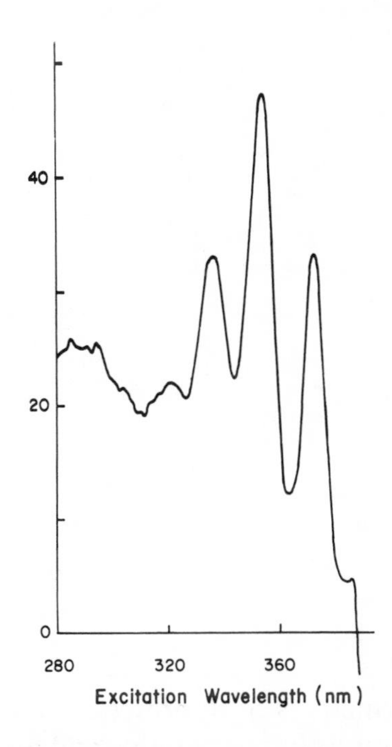

Fig. 3.6c. *The excitation spectrum of anthracene. [The emission wavelength is 400 nm.]*

SAQ 3.6g

Draw up a list of all the factors which contribute to the blank in a fluorescence assay at low concentrations.

SAQ 3.6g

So much for the principles of fluorescence measurement and their potential and limitations in quantitative and qualitative analysis. We have covered the ground quite thoroughly and you should now be aware that fluorescence can be used in a number of different ways with low limits of detection as its chief advantage. We are now ready to explore how it is applied in practice and the types of analyte for which it is most commonly used. This will be the subject of Part 4 of this Unit.

SUMMARY AND OBJECTIVES

Summary

The intensity of fluorescence I_f from a solution containing an analyte of molecular absorptivity ϵ at a concentration c is given by

$$I_f = \phi_f I_o e^{-2.303\epsilon cd}$$

At low concentrations, where $2.303\epsilon cd < 0.05$ this expression simplifies to $I_f = 2.303\phi_f I_0 \epsilon cd$ and a linear calibration graph is obtained in fluorimetric determinations. At higher concentrations where the absorbance of the solution exceeds 0.02, the graph becomes curved because of the influence of higher terms in the expansion of $e^{-2.303\epsilon cd}$. The concentration at the limit of linearity is only about an order of magnitude greater than the detection limit in absorption spectroscopy. The detection limit in spectrofluorimetry is some 1000 times lower and the dynamic range some 100 times greater.

The low concentrations encountered in fluorescence work are often quoted in units of parts per billion (ppb). However, this unit is ambiguous and appropriate *w/v* units such as $\mu g\ dm^{-3}$ are to be preferred.

Fluorescence calibration graphs constructed for higher concentrations where absorbance exceeds 0.03 show additional curvature beyond that caused by higher terms in the expansion of $e^{-2.303\epsilon cd}$. This is due to the inner filter effect. When 90° geometry is used the curve eventually passes through a maximum because the incident radiation does not penetrate beyond the first mm or so of sample and the fluorescence is emitted outside the region viewed by the emission monochromator. The shape of both excitation and emission bands becomes distorted by the inner filter effect. The effect can also arise if absorbing species other than the analyte itself are present. The consequences of the inner filter effect can be reduced by using frontal illumination or straight-through geometry.

The intensity of fluorescence may be reduced by the presence of molecules which can remove the energy from excited analyte molecules before they can emit radiation. Common examples of species which can quench fluorescence in this way are dissolved oxygen, halogen ions, and compounds containing heavy atoms. At high concentrations, quenching is often due to molecules of the analyte itself.

Many organic molecules decompose under the conditions of irradiation with intense uv radiation necessary to excite fluorescence. The fluorescence of the analyte may then decrease with time or, if one of the products of photodecomposition is itself fluorescent, the background emission will increase. In these circumstances it is advantageous to excite the fluorescence at the longest possible wavelength and to keep exposure of the sample to the radiation to a minimum. Dissolved oxygen often participates in photochemical reactions as well as being a quenching agent and it then becomes necessary to remove it by bubbling nitrogen through samples before measurement.

The sensitivity of fluorimetric methods of analysis is governed by three main factors – the absolute sensitivity of the analyte as determined by its absorptivity and fluorescence efficiency, the instrumental sensitivity determined by the source intensity, detector sensitivity and the optical efficiency of the instrument, and the method sensitivity determined largely by the limitation imposed by the blank. Instrumental sensitivity is often quoted in terms of the detection limit for quinine under standard conditions though the signal/noise ratio for the Raman band at 350 nm has recently been adopted as a more satisfactory standard.

The blank reading in a fluorimetric determination is of major importance since it frequently prevents its use down to low concentration which appear feasible from signal/noise measurements on solutions of pure analyte. Principal contributions to the blank are fluorescent impurities from various sources, suspended particles and, ultimately, the Rayleigh and Raman scattering of the solvent. Both types of scattering are of narrow band-width compared with the emission band of the analyte and their interference can be minimised by operating with as high an instrumental band-pass as possible subject to other interferences and stray light considerations. For low level trace analysis it is necessary to take stringent precautions in the selection of reagents, the preparation of solutions and the storage and handling of glassware.

Objectives

You should now be able to:

- recognise that the analytical status of fluorescence spectroscopy is essentially similar to that of uv/visible absorption spectroscopy with enhanced sensitivity, dynamic range and specificity, but reduced precision;

- derive the relationship between fluorescence intensity and concentration;
- distinguish between the linear region of the fluorescence calibration and the region where curvature arises due to higher powers in the expansion of the exponential term in concentration;
- explain the inner filter effect and to recognise the influence of high concentrations of both analyte and other species in causing addition curvature of calibration graphs and distortion of fluorescence bands;
- recognise that the overall sensitivity of a fluorescence method includes the absolute sensitivity for the analyte, instrumental sensitivity and factors due to the method used to prepare the sample and to identify these components;
- identify the factors which contribute to the blank in fluorescence measurements and take steps to minimise them;
- distinguish between Rayleigh and Raman scattering and their effects on fluorescence measurements;
- describe the simpler processes leading to fluorescence quenching and recognise molecules containing heavy atoms or unpaired electrons as potential quenching agents;
- recognise that quenching can be used as a method of analysis for the quenching species;
- recognise that dissolved oxygen is a common quenching agent and often induces photodecomposition;
- describe the effects of photodecomposition on fluorescence measurements, recognise the problem when it arises in a practical situation and take steps to minimise it.

4. Photoluminescence Methods in Analysis

4.1. INTRODUCTION

There are many ways in which fluorescence can be used in analysis but most of the routine applications fall into one of the following three groups:

1. Direct methods – in which the natural fluorescence of the analyte molecule itself is measured;

2. Derivatisation methods – in which the analyte is non-fluorescent but is converted into a fluorescent derivative.

3. Quenching methods – in which the analytical signal is the reduction in the intensity of some fluorescent species due to the quenching action of the analyte.

Each of these groups can be further sub-divided according to the chemical nature of the analyte or the experimental method used to excite the luminescence. In Part 4 we shall review each of these groups in turn and look at some typical examples. We shall also discuss the use of scanning techniques in analytical work. In Part 5 we shall look at some more specialised techniques (with an eye to future developments in some cases) and also illustrate the practical aspects of fluorimetric analysis by way of some selected case studies.

4.2. DIRECT METHODS

Within this category the main subdivision is between inorganic and organic species though the latter is by far the larger.

4.2.1. Inorganic Species

Although the list of inorganic species of analytical interest is endless, the number which are fluorescent is exceedingly limited. The only simple species which are naturally fluorescent in solution are the ions of the lanthanide (rare earth) and actinide (trans-uranic) elements and one or two odd cases such as the thallous ion in the presence of excess chloride ion and tin in sulphuric acid.

Π What feature is common to the atomic structure of the lanthanide and actinide elements?

The common feature is the presence of an incomplete shell of f electrons which can be promoted by the absorption of uv radiation to orbitals of higher energy. (A similar situation involving d electrons accounts for the colours of transition metal ions.)

Π As a group, the lanthanides are not very well known to students but you may have come across one or two of them which have specific scientific applications. Can you name any of them?

The two commonest are probably cerium and europium. The complete list is given in Fig. 4.2a together with the number of 4f electrons to be added to the xenon core to give the electronic configuration of the trivalent ions, M^{3+}, the form in which they are normally encountered in aqueous solution.

Most of the elements are fluorescent in the solid state but only the middle five, Sm, Eu, Gd, Tb, and Dy, are fluorescent in aqueous solution.

Π Why do you think the lanthanides are fluorescent? How do the principles discussed in Part 1 apply?

[Xe core:
$1s^2$ $2s^2$ $2p^6$ $3s^2$ $3p^6$ $3d^{10}$ $4s^2$ $4p^6$ $4d^{10}$ $5s^2$ $5p^6$]

		4f
Cerium	Ce	1
Praseodymium	Pr	2
Neodymium	Nd	3
Promethium	Pm	4
Samarium	Sm	5
Europium	Eu	6
Gadolinium	Gd	7
Terbium	Tb	8
Dysprosium	Dy	9
Holmium	Ho	10
Erbium	Er	11
Thulium	Tm	12
Ytterbium	Yb	13
Lutecium	Lu	14

Fig. 4.2a. *Lanthanides with the number of f electrons present in the M^{3+} ion*

You should have been able to give the trivial, stock answer that the excited states are not easily deactivated by internal conversion or collision. To explain why these radiationless transfer mechanisms are inefficient is a little more tricky but it must be related to the properties of the 4f electrons. The fact is that they lie inside the region of the atom occupied by the 5s and 5p electrons and are screened by them from external influences. Consequently, transitions between the 4f orbitals are very little affected by the environment in which the atom is placed and give emission spectra consisting of sharp lines more characteristic of atomic spectra than molecular spectra. In contrast to organic spectra, more than one electronic transition is involved.

The direct fluorescence emission of the lanthanides is in fact rather weak and, though it is of very great academic interest, its analytical importance is negligible. On the other hand, certain of the lanthanides are becoming of particular interest to analysts because of their rather special spectroscopic properties and we shall meet them again later in the context of fluorescent labels for organic compounds.

The actinides are of much greater commercial and therefore analytical importance though analysis is usually carried out by radiochemical methods. However, spectrofluorimetric methods are frequently used for the commonest of these elements, uranium. The most important species in aqueous solution is the uranyl(VI) ion, UO_2^{2+}, which exhibits a green fluorescence excited by blue incident radiation. Because it is a molecular species the spectrum shows a pronounced vibrational fine structure and the appearance of the band differs very little in different salts. The intensity varies widely with different anions and is particularly enhanced by the presence of sulphuric acid as shown in Fig. 4.2b.

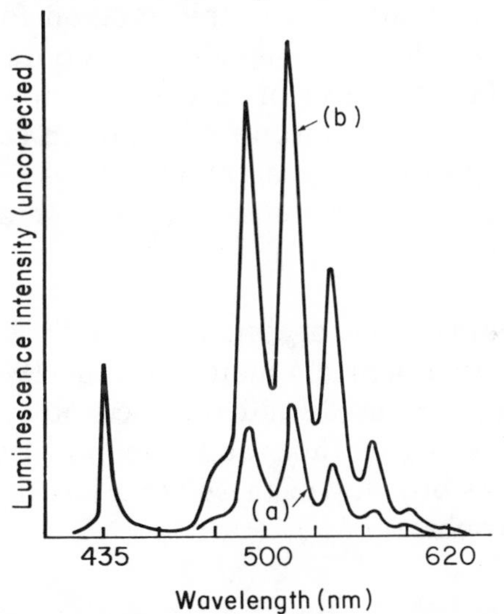

Fig. 4.2b. *Fluorescence spectrum of uranyl sulphate (5×10^{-3} mol dm^{-3}) in*

(*a*) Sulphuric acid (0.05 mol dm^{-3})
(*b*) Sulphuric acid (2.50 mol dm^{-3})
[Excitation 436 nm, band-pass 40 nm]

Like the lanthanides, the sensitivity which can be achieved using the fluorescence of the simple ion is poor and in practice derivatisation methods are preferred.

Π In the solid state the range of photoluminescent inorganic compounds is very much wider and both fluorescence and phosphorescence emission is commonly encountered. We have already mentioned two inorganic solids which fluoresce in the early parts of this unit. What are they? (One of them gave its name to the phenomenon).

Fluorspar, the mineral form of calcium fluoride, is the one from which the word 'fluorescence' is derived. The other is quartz though, as we mentioned in the earlier discussion, it is an impurity (aluminium) in the quartz, not the silica itself which is responsible for the fluorescence. This is a common feature of solid luminescent inorganic compounds and many minerals are fluorescent where their synthetic equivalents prepared from pure starting materials are not. This provides a simple method for distinguishing between real and synthetic (or fake!) gem stones – ruby is basically alumina containing traces of chromium impurity. Qualitative analysis of this type is straightforward and often carried out by visual examination by experts. Quantitative analysis is more difficult because of the problems encountered quite generally in analysing materials in the solid state.

It is worth mentioning that inorganic luminescent materials such as zinc sulphide, to which small amounts (less than 1%) of metals such as Cu, Mn, and the actinides are added as 'activators', are used in the picture screens of television tubes and video display units. Because of the long lifetimes of many of these materials they are generally referred to as 'phosphors'.

4.2.2. Organic Compounds

The range of organic compounds which can be directly excited in solution to emit fluorescence is very much wider than in the inorganic field. We have already discussed in Part 1 the types of organic compound which are likely to be fluorescent and the factors determining their fluorescence efficiency. We now need to extend this discussion and consider the areas of analysis where direct fluorescence provides a viable method.

Π Let's start however by seeing whether you can recall some of the principles set out in Part 1. What type of compound would you expect to have a high fluorescence efficiency?

The usual requirement is for a large molecule containing a conjugated system of some sort which has a rigid structure due to ring formation. It will also have a high absorptivity. The fluorescence efficiency will be reduced by the presence of atoms carrying lone pairs where the $n \rightarrow \pi^*$ transition gives rise to the lowest excited state of the molecule. The presence of heavy atoms will also reduce the fluorescence efficiency.

Π Give two examples of compounds which fluoresce and one where the efficiency is reduced by structural features.

Your examples might well have included anthracene and other polycyclic aromatic hydrocarbons, fluorescein, rhodamine, tryptophan, etc. whose structures are given in Part 1. Compounds whose fluorescence is reduced by the presence of groups containing lone pairs include carbonyl compounds such as benzophenone, benzoic acid and acid derivatives such as esters and amides generally. Many nitrogen compounds such as pyridine also come into this category. The commonest 'heavy atoms' are bromine and iodine so that compounds such as bromobenzene are not fluorescent. If you were unable to recall these examples you should re-read the relevant section (1.4) of Part 1. Of course, the reduction of fluorescence intensity in these compounds may well be offset by an increase in the phosphorescence efficiency because the triplet state becomes appreciably populated by inter-system crossing. So, when fluorescence fails, the compounds may still be accessible by phosphorescence techniques where these are available.

The scope of natural fluorescence is indeed quite wide and methods are commonly employed for the following compounds or groups of compounds:

(*a*) Polyaromatic hydrocarbons (PAH), particularly the higher members such as benzpyrene, which are present in the tar from tobacco smoke and have a powerful carcinogenic action. They are also found in traces in rivers after forest fires, having been washed out of the atmosphere by subsequent rainfall.

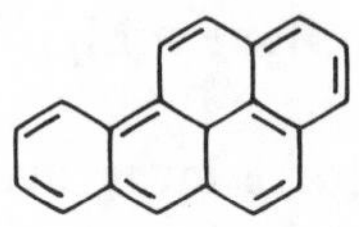

(*b*) Vitamin A, which has a blue-green fluorescence and an excitation maximum of about 330 nm. The most significant feature of the structure of this molecule is the conjugated chain of 5 double bonds

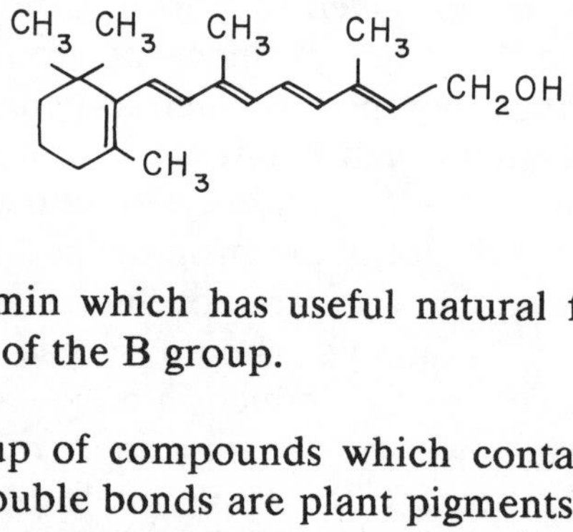

Another vitamin which has useful natural fluorescence is riboflavin, one of the B group.

(*c*) Another group of compounds which contain long chains of conjugated double bonds are plant pigments including the anthocyanins, lycopene and the carotenes (open chains like vitamin A) and the porphyrins (condensed ring structures like chlorophyll).

(*d*) The fluorescence of quinine sulphate has frequently been mentioned. This compound is readily available and its excitation and emission wavelength fall in the middle of the uv/visible region so that it is often used as a test material to check the performance of fluorescence instruments. From the point of view of analysis, however, quinine is an example of a drug which can be determined by fluorescence spectroscopy. In fact, the estimation of quinine in tonic water is a routine analysis in the soft drinks industry.

(*e*) Many important compounds present in the body fluids are fluorescent when dissolved in concentrated sulphuric acid. These include steroids such as cortisol and cholesterol and a number of hormones. All the steroid molecules contain fused ring systems as shown below. Although this is often described as 'natural fluorescence' it is likely that the sulphuric acid is not merely a solvent but causes condensation reaction between and within these molecules. In fact, the precise nature of the fluorescent species has not been definitely established.

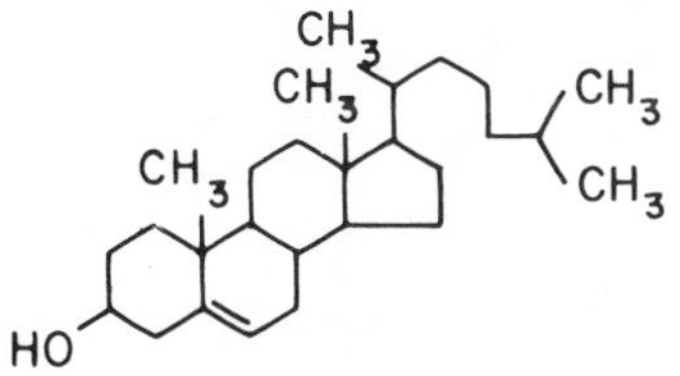

Cholesterol

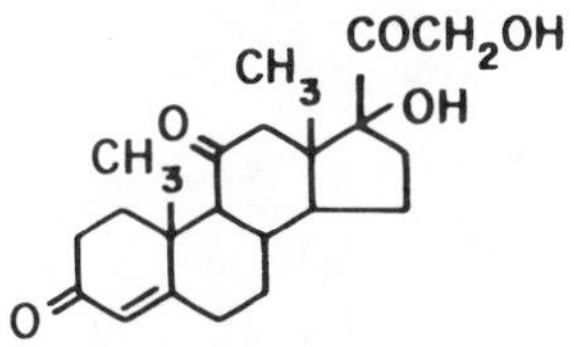

Cortisone

Apart from the compounds mentioned above there is a very large number of compounds of analytical importance which exhibit natural fluorescence. However, most of these emit at relatively short wavelengths in the uv region where there is great difficulty in avoiding interference between them. Consequently it is customary to use derivatisation methods to improve selectivity either by selective reaction between the analyte and the reagent or by shifting the fluorescence to a region where there is less interference. *In practice, therefore, the usefulness of natural fluorescence is limited to those compounds which fluoresce at long wavelengths, well clear of the fluorescence of other compounds which occur with them.*

SAQ 4.2a

Which of the following lanthanides is fluorescent in solution?

Eu Lu Ce Tb

SAQ 4.2a

SAQ 4.2b

(*a*) What simple uranium-containing species is fluorescent?

(*b*) What colour is the fluorescence?

(*c*) Under what conditions is it enhanced?

(*d*) Why does its emission band show fine structure?

SAQ 4.2c

Which of the following compounds could be analysed using their natural fluorescence?

tryptophan
vitamin A
ethanedicarboxylic acid
cortisol

4.3. DERIVATISATION METHODS

Perhaps the simplest example of a fluorescent derivative is the complex formed between a metal ion and a donor organic molecule. Ideally both of these components should be non-fluorescent so that when the fluorescent complex is formed its fluorescence can be observed easily. In practice any natural fluorescence of the components can be tolerated provided that it is weak and occurs at a wavelength well removed from that of the complex. This method is entirely analogous to the formation of coloured complexes between metal ions and colorimetric reagents which are colourless or absorb at different wavelengths from the complex. In fact, many of the reagents used for specific analytes in spectrophotometry produce complexes which are also fluorescent and the use of fluorescence measurements rather than absorption measurements may well be advantageous in certain analyses.

Π When would you suppose the fluorescence techniques would offer advantages over the spectrophotometric technique?

At very low concentrations of analyte which are below the limit for colorimetry. Also where there is a problem of interference in spectrophotometry which could be avoided in spectrofluorimetry because of the greater selectivity made possible because there are two analytical wavelengths.

Complex formation can be used in two senses – to determine either the metal species where the organic molecule is the reagent or the organic compound where the metal ion is the reagent. The former is the more common but there is currently much research interest in the use of metals, particularly the lanthanides, as 'probes' to determine organic species in quite complex matrices, often using fluid phosphorescence techniques. These methods are likely to become important in some areas of analytical work in the future.

4.3.1. Fluorimetric Reagents for Metals

The most effective type of complexing reagent for the formation of a fluorescent complex is one with which the metal forms bonds in two positions to produce a chelate ring. This normally requires the presence of two functional groups in the reagents, one containing an acidic (replaceable) hydrogen atom and the other an atom carrying a lone pair. The acidic hydrogen is usually supplied by an —OH group (phenolic or carboxylic) and the lone pair comes from an oxygen, nitrogen or sulphur atom. For stable complexes to be formed the two groups must be situated so that chelation with the metal produces a

5- or 6-membered ring. A typical example is 2-hydroxybenzaldehyde (salicylaldehyde) which forms complexes with metals as shown here.

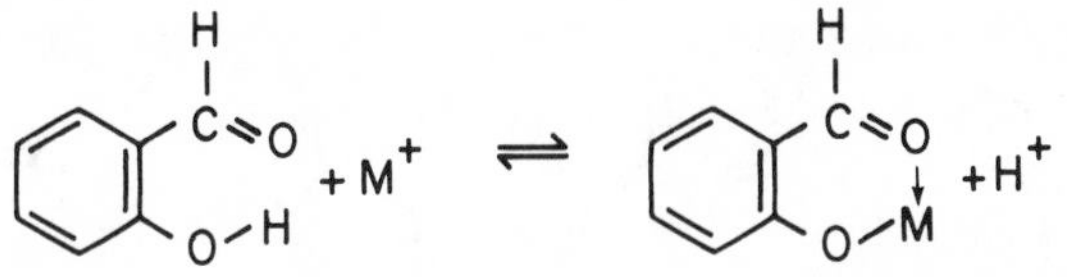

Π There are two features of this process which favour the formation of a complex which will be fluorescent. What are they?

The formation of a ring with the metal atom increases the rigidity of the molecule and effectively increases the size of the chromophore and hence the absorptivity.

The donation of the lone pair to the metal removes the possibility of a low-lying $\rightarrow \pi^*$ excited state which may have caused the reagent itself to be non-fluorescent.

Another very well-known complexing agent which forms complexes with a large number of metals is 8-hydroxyquinoline ('oxine')

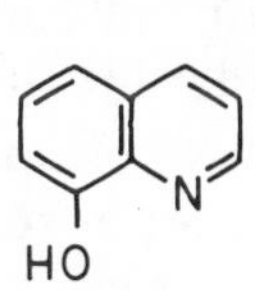

This is used as a reagent in several branches of analytical chemistry, including gravimetry and colorimetry, and many of the complexes are fluorescent. The sulphonic acid group, SO_3H, is often incorporated at the 5- position to facilitate the use of the reagent in aqueous solution.

Π Draw the structures of the complexes formed between oxine and Mg^{2+} and Al^{3+} which are planar and octahedral respectively. What is the ring size in these complexes and what is the charge on each of them?

The actual structures are

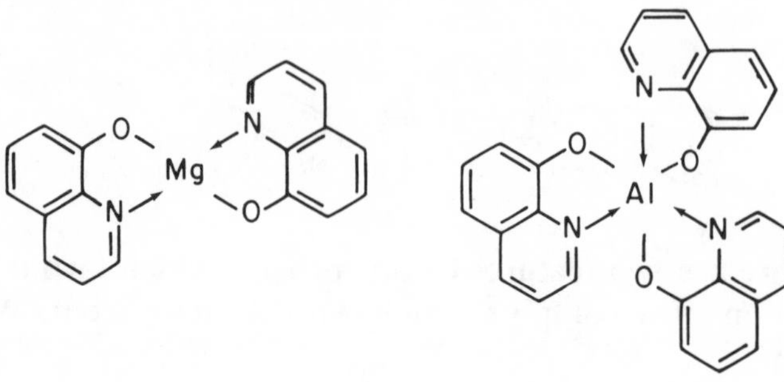

In oxine complexes the ring is 5-membered. In the case of both magnesium and aluminium the complex carries no charge. When a neutral complex is formed it is possible to extract it into an organic solvent such as ether or chloroform. This is a very useful procedure since it enables us to separate the complex from charged complexes of other metals and other ions which might upset the analysis. This is one way in which complex formation can be made selective. In other cases, the reagent can be made more selective by control of pH or steric factors.

A great deal of research has been carried out to design reagents which are specific for a particular metal or group of metals. Much of this work was directed towards developing reagents for spectrophotometry but, as we have seen, these reagents are frequently useful in fluorimetry as well. The literature now contains descriptions of thousands of complexing agents and analytical methods based on them. Fortunately, references to most of the important ones covering a period of at least 30 years are collected together in review articles and specialist text-books so that, except for the most recent methods, it is relatively easy to find information on the complexes of any particular analyte.

The structures of a few typical complexing agents are shown below and some examples of fluorescent complexes and their properties are given in Fig. 4.3a.

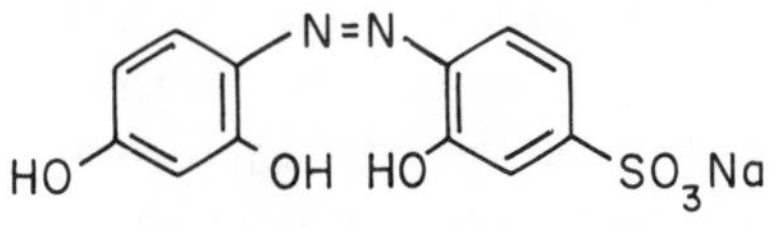

Acid alizarin garnet red

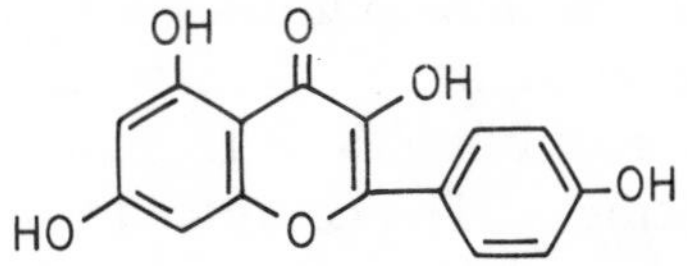

Morin

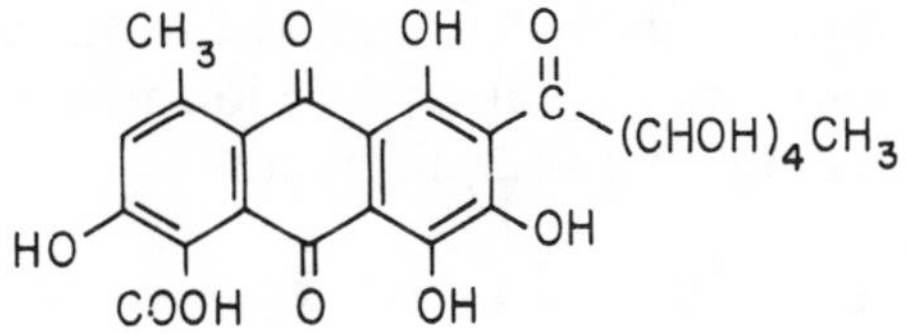

Carminic acid

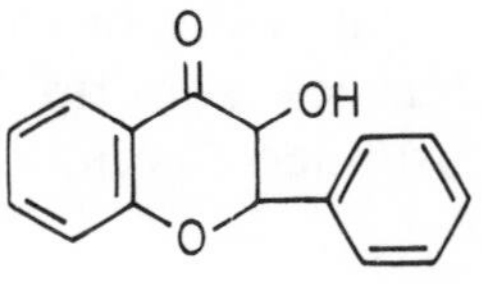

3-hydroxy flavone

Metal	Reagent	λ_{ex} (nm)	λ_{em} (nm)	Limit of Detection ($mg\ dm^{-3}$)
Aluminium	Alizarin garnet red	470	580	0.007
Beryllium	Morin	420	525	0.01
Magnesium	8-Hydroxyquinoline	420	530	0.01
Manganese	Carminic acid	467	556	0.9
Tin	3-Hydroxyflavone	415	495	0.004
Zinc	8-Quinolinol	375	517	0.5

Fig. 4.3a. *Typical fluorescent complexes*

4.3.2. Inorganic 'Probes' for Organic Analysis

Although it is possible to reverse the role of 'analyte' and 'reagent' in complex formation so that the organic component can be determined using the metal ion as the reagent, the highly specialised organic reagents for metals are not likely to turn up in everyday analysis. However, it is possible that some related, more mundane compound might be determined in this way and the metal ion used in these circumstances is referred to as a 'probe'. This technique is potentially very useful and much research interest has been aroused with the aim of finding metal ions which will form a complex with one particular analyte in the presence of other species with the minimum of operations to control the conditions to achieve the desired specificity. The complex formed must have very characteristic properties so that it can be identified and measured easily even when the sample produces a high fluorescent background.

Π What is the most important property that will avoid interference from the background? What type of ion is likely to produce a complex meeting these requirements?

The fluorescence emission should occur at long wavelength, preferably above 500 nm, since background emission is most severe below 350 nm. Complexes of the lanthanides have very characteristic 'line' emission which usually occurs in the visible region

Europium and terbium have been found to show the greatest promise and useful complexes are formed with proteins and other compounds of biological interest. This is of particular significance because the analysis of samples of biological origin is plagued by problems of interference and by high blank values. As with other lanthanide complexes, the emission spectrum is virtually identical with that of the free ion but the intensity is increased by a factor of 10^5 in the case of terbium.

Π What disadvantage results from this observation?

Since all complexes of terbium give the same emission spectrum as the free ion, if more than one complex is formed it will not be possible to discriminate between them by varying the wavelength at which the emission is measured. There are two other variables that we can make use of before we have to resort to the inconvenience of chemical treatment or separation. One arises from the fact that the lifetime of the emission is of the order of several milliseconds.

Π How can we make use of this property and what is the other variable which might prove effective in making the analysis more specific?

We could use time resolution if we had access to a suitably equipped instrument. This would enable us to delay the observation until the emission had decayed from all the species except the one of interest, provided of course that it had the longest lifetime. This certainly works in the case of the free ion since it invariably has a shorter lifetime than the complex. However, since the intensity is so much weaker, the emission from the free ion is not really a serious problem. The other variable we might employ is the excitation wavelength which is characteristic of the organic part of the molecule.

Although the emission from the terbium/protein complex has a long lifetime, it is not strictly phosphorescence because it does not originate from a triplet state. The fact that the emission consists of a line spectrum which is essentially the same for all complexes and for the free ion indicates that the transitions concerned occur between the electronic levels of the atom. With complexes of main group elements, the emission is associated with transition of electrons located predominantly within the organic part of the molecule so that a typical molecular band spectrum is observed. Complexes of the transition metals are intermediate in their behaviour though few of them are fluorescent. When they are, the fluorescence efficiency is always very low and the emission may be atomic or molecular. For example, the oxine complex of cobalt has a fluorescence efficiency of 0.01 and emits a band spectrum while the oxine complex of chromium has a fluorescence efficiency of 0.0033 and the fluorescence is 'atomic' in nature. The fluorescence efficiencies of iron and nickel oxine are more than an order of magnitude lower.

The mechanism of these processes are shown in Fig. 4.3b which may help to rationalise the fluorescence behaviour of metal complexes. However, don't worry if you find energy level diagrams difficult – but you will have to remember the facts!

In the following diagrams, the levels in which the promoted electron is in an atomic orbital are labelled a_1 and a_2. S_1 and T_1 are normal singlet and triplet excited states of the complex.

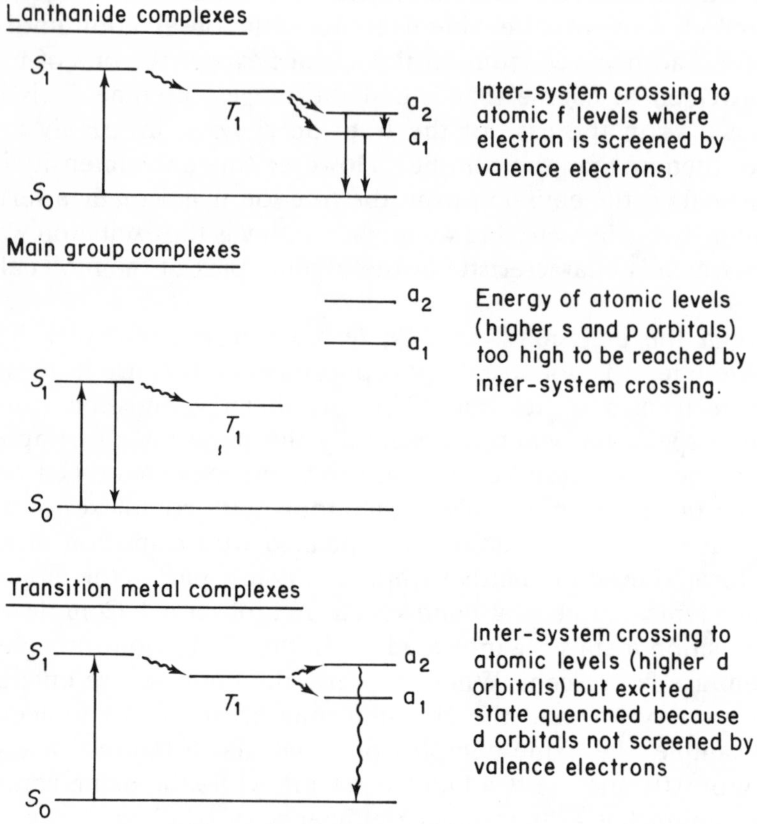

Lanthanide complexes

S_1 T_1 a_2 a_1 S_0

Inter-system crossing to atomic f levels where electron is screened by valence electrons.

Main group complexes

a_2 a_1 S_1 T_1 S_0

Energy of atomic levels (higher s and p orbitals) too high to be reached by inter-system crossing.

Transition metal complexes

S_1 T_1 a_2 a_1 S_0

Inter-system crossing to atomic levels (higher d orbitals) but excited state quenched because d orbitals not screened by valence electrons

Fig. 4.3b. *Energy level diagrams for metal complexes*

Π Would you expect there to be a mirror-image relationship between the emission and excitation spectra of lanthanide complexes?

No. The excitation spectrum would be a band spectrum characteristic of the organic part of the complex. With protein complexes it often resembles the excitation spectrum of tryptophan which is commonly the only amino acid residue in natural proteins which absorbs in the uv region.

4.3.3. Derivatives of Non-metals and Anions

Non-metallic elements and anionic species are frequently something of a problem because they are not directly accessible by atomic spectroscopy, which is the traditional method of analysis for metals, and they do not readily form donor complexes which provide an alternative method either by spectrophotometry or spectrofluorimetry. The most well-known fluorimetric methods are those for boron and selenium which both involve condensation reactions leading to ring closure, and some special specific methods for ions such as cyanide, sulphide, fluoride, and phosphate.

The most well-established and carefully studied derivative of boron is the condensation product formed between boric acid (the most common boron compound encountered in analysis) and benzoin.

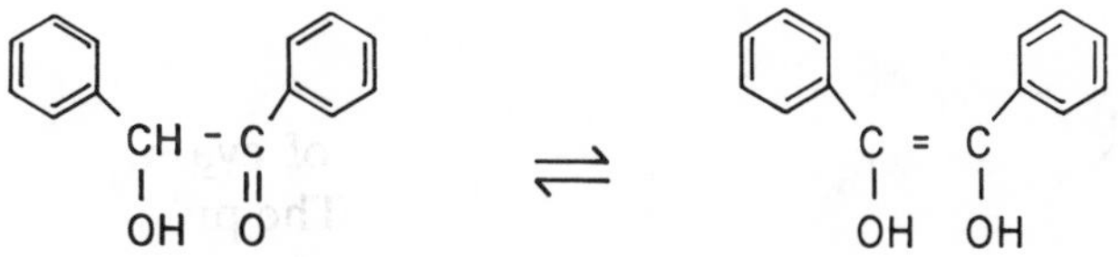

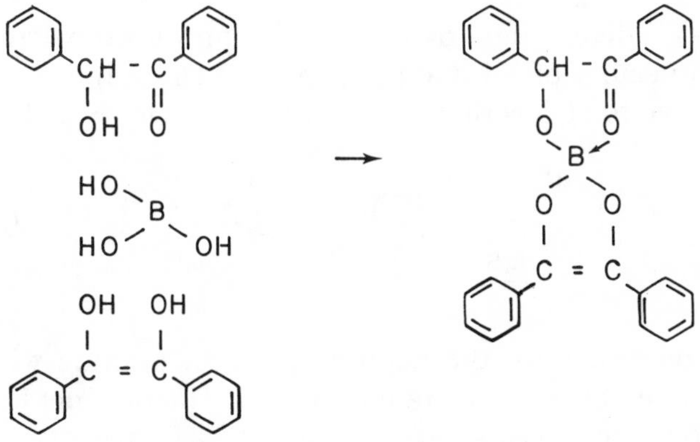

A similar reaction occurs with 2,2′-dihydroxybenzophenone. Draw the structure for this derivative. (You should be able to deduce the structure of the reagent from its systematic name.)

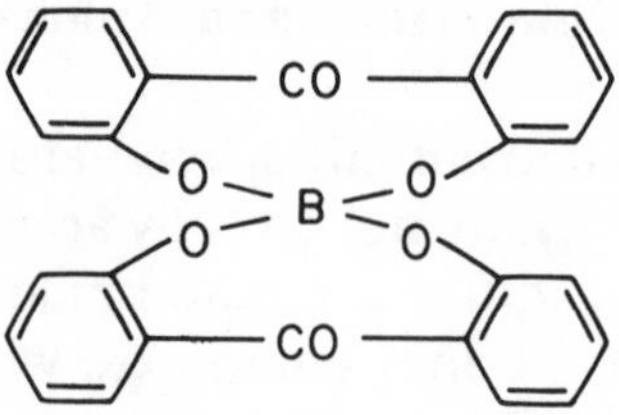

Note that all boron complexes are tetrahedral.

Π How does the benzoin/boric acid reaction differ from the complexing reactions used for metals?

It involves the elimination of water between hydroxyl groups. The method for selenium (as selenious acid) is also well-established and involves a ring-closure reaction with 'DAN' (2,3-diaminonaphthalene), again with the elimination of water.

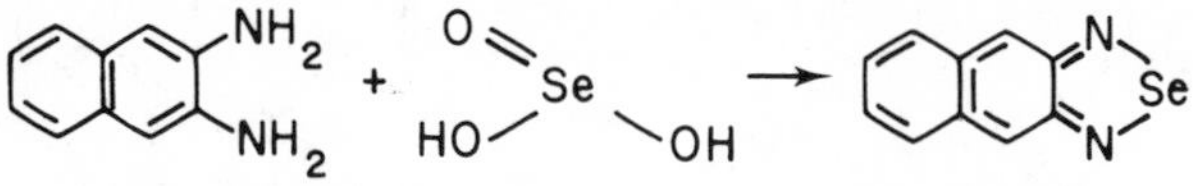

4,5-benzopiazselenol

Sulphur too is determined by a derivative produced by ring closure with two molecules of 1,4-diaminobenzene. The sulphur compound is first reduced to H_2S with hypophosphorous acid + hydriodic acid

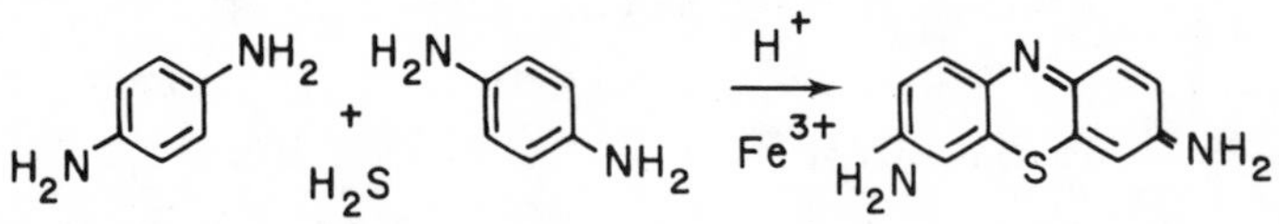

Of the anion reactions, the determination of cyanide with benzoquinone is one of the most straightforward. The product has a blue–green fluorescence (λ_{em} = 480 nm, λ_{ex} = 400 nm).

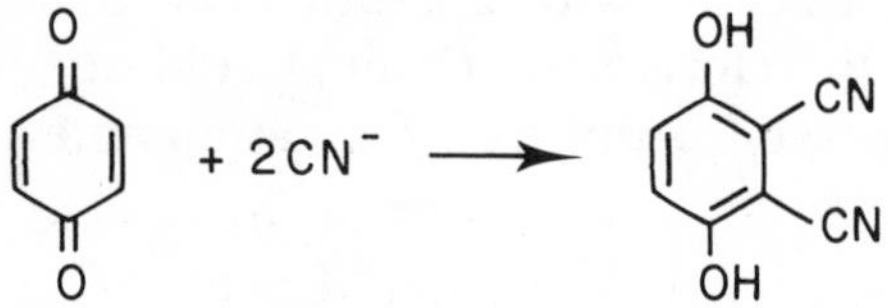

In the case of fluoride many methods have been proposed but the most effective involves the formation of a ternary complex with zirconium and calcium blue at pH 2.5 which has a blue fluorescence (λ_{em} = 410 nm, λ_{ex} = 350 nm).

The phosphate determination starts like the spectrophotometric method with the formation of the phosphomolybdate ion, but, instead of reduction, it is treated with rhodamine B to form an ion association complex which is extracted into chloroform/butanol (4 : 1) in which it gives a yellow fluorescence (λ_{em} = 575 nm, λ_{ex} = 350 nm).

4.3.4. Organic Derivatisation

We saw in Section 4.2.2 that, whilst a considerable number of naturally fluorescent organic compounds can be analysed by direct methods, problems arise when the emission is of short wavelength (below 300 nm) because of serious interference from the wide range of fluorescent compounds present at trace levels in the environment. The development of techniques based on time resolution where longer lived photoluminescence can be produced has extended the range of application of direct methods. However the most significant increase in the number of materials which can be analysed by fluorescence measurement has been achieved by forming derivatives which fluoresce at wavelengths well above the range where interferences are a problem. This can be applied to compounds whose natural fluorescence is at short wavelength or to compounds which do not fluoresce at all. Although inorganic 'probes' are being developed for some classes of organic analytes, most organic derivatisation involves the use of purely organic reagents.

Π How would the electronic structure of a fluorescent derivative be likely to differ from that of the analyte species or the reagent used to make the derivative?

The derivative is likely to have an enlarged π-system compared with the analyte so that it absorbs and emits at longer wavelength. In some cases the reagent supplies the basic π-system which is then enlarged during the reaction forming the derivative so that the fluorescence is well clear of any emission from the reagent itself. Another feature that would enhance the fluorescence of the derivative would be the formation of ring systems to increase the rigidity of the molecule.

Π What type of reaction is most likely to be used for the formation of a fluorescent derivative?

The analyte and the reagent have got to be linked together, usually by some sort of condensation reaction in which small molecules such as water or HCl are removed from the two molecules.

$$\text{NH-NH}_2 + \text{O=C(H)} \xrightarrow{-H_2O} \text{NH-N=CH}$$

$$\text{SO}_2\text{Cl} + \text{H}_2\text{N-CH(COOH)-R'} \xrightarrow{-HCl} \text{SO}_2\text{-NH-CH(COOH)-R'}$$

In some cases the reaction may well produce an extra double-bond which extends the π-system already present

$$\text{C(H)=O} + \text{H}_2\text{N} \xrightarrow{-H_2O} \text{CH=N}$$

To achieve ring closure a reagent is required with reactive groups situated at appropriate points in the molecule – rather like the relative position of groups in a chelating agent. Disubstituted aromatic compounds such as 1,3-dihydroxybenzene (resorcinol) and 1,2-diformylbenzene (phthalaldehyde)

OH, OH; CHO, CHO

are often used as the reagent which condenses with the analyte to form a ring.

eg resorcinol with polycarboxylic acids

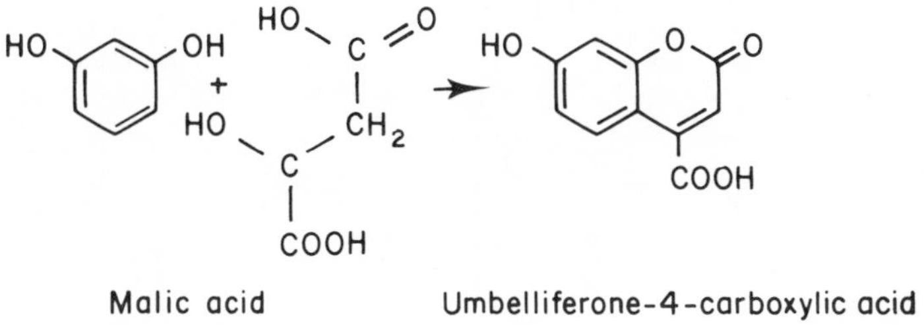

Malic acid Umbelliferone-4-carboxylic acid

In some cases the analyte may already have suitable groups in positions for ring closure with appropriate chemical treatment.

But – we are getting too deeply involved in organic chemistry! Obviously the development of these procedures is going to require some sophisticated organic chemistry from somebody, but, for the method to be of general application it must be reliable, rapid and quantitative with no complicating by-products. By its very nature it is likely to be specific – indeed it is likely that methods will have to be developed for each different analyte or type of analyte. Ideally these should be as simple as the complexing reactions used for metals.

SAQ 4.3a

Draw the structure of the complex formed between boric acid and 1,2-dihydroxybenzene. Why should it be fluorescent?

SAQ 4.3a

SAQ 4.3b

Draw the structure of the complex formed between tin and 3-hydroxyflavone assuming that it is octahedral (see Fig. 4.3a).

SAQ 4.3c

Are the following statements about complexes of the lanthanides true or false?

(*i*) The excited state is more easily deactivated than in transition metal complexes.

(*ii*) The emission spectrum is a mirror image of the excitation spectrum.

(*iii*) The emission spectrum consists of a number of very sharp bands.

(*iv*) The lifetime of the emission is greater than 1 μs.

(*v*) The emission can be described as phosphorescence.

(*vi*) The emission can be observed with the sample in the liquid state.

SAQ 4.3d

State whether completion of the following sentence by the phrases (*i*) to (*v*) gives a statement which is true or false:

Derivatisation of organic compounds

(*i*) shifts any natural fluorescence to longer wavelength.

(*ii*) increases the size of the π-system.

(*iii*) involves elimination of small molecules like H_2O and HCl.

(*iv*) gives a product with a higher fluorescence efficiency.

(*v*) is specific because a reagent reacts only with one class of compound.

SAQ 4.3e

State *three* ways in which the emission from a lanthanide complex differs from phosphorescence.

SAQ 4.3f

List the main factors which can be used to achieve higher selectivity in a fluorimetric analysis.

SAQ 4.3f

4.4. QUENCHING METHODS

The principle of a quenching method is that a fluorescent species has to be found whose fluorescence emission is quenched by the analyte. The fluorescence intensity therefore falls as the concentration of analyte increases.

Π Compared with the normal fluorescence technique already discussed quenching suffers from two major disadvantages. Perhaps you can see what these are. Try to describe them.

The main disadvantage is that quenching is completely non-specific. There is no measurable parameter which is uniquely related to the analyte – like the appearance of the spectra of fluorescent species in the normal techniques. Consequently the applications of quenching methods are limited to analyses where the analyte is known to be the only species capable of quenching the fluorescence – a very severe limitation.

The other disadvantage is that the measured parameter, the fluorescence intensity, decreases as the concentration increases. Furthermore, the calibration is also non-linear.

Π A simple example of quenching is the determination of halide ions which quench the fluorescence of quinine at quite low concentrations. Some actual calibration graphs are shown in Fig. 4.4a. How does the performance of the method vary for the different halides?

The efficiency of the quenching increases with the mass of the halogen so the method is far more sensitive for iodide than for chloride.

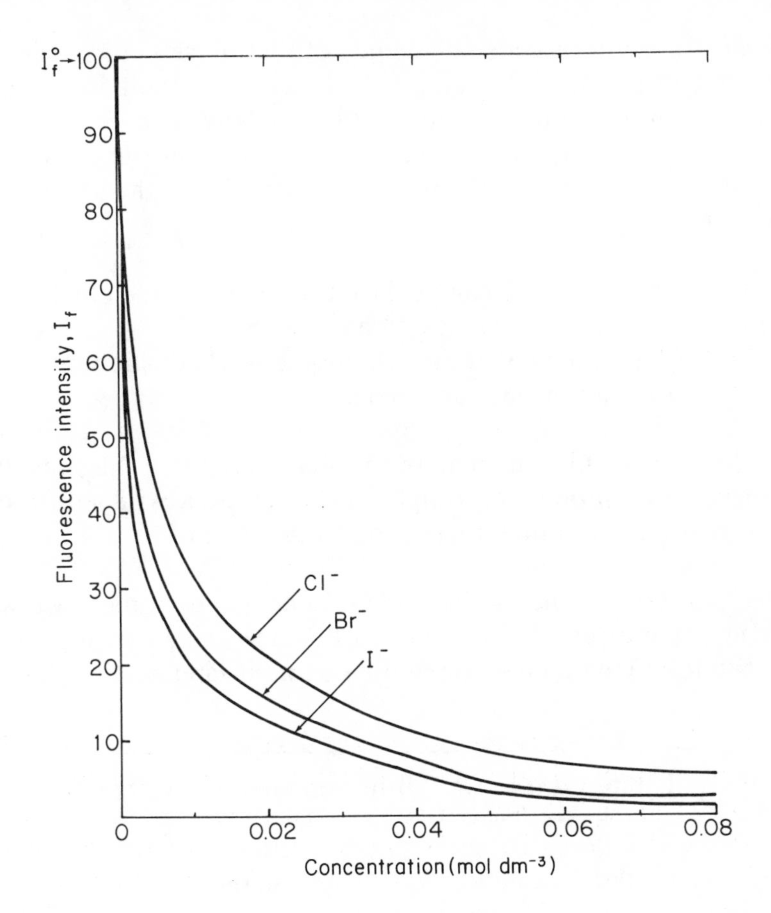

Fig. 4.4a. *Quenching of quinine by halide ions at different concentrations*

It is possible to plot the data from a quenching method in a way which produces a straight line calibration by making use of a very long established mathematical treatment first published by Stern and Volmer. The outcome of this treatment (which we do not need to follow through) is expressed in the Stern–Volmer equation

$$\frac{\phi_f^o}{\phi_f} - 1 = Kc$$

where ϕ_f^o is the fluorescence efficiency of the fluorescent species in the absence of the quencher and ϕ_f is the value of the fluorescence efficiency when the concentration of the quencher is c. K is known as the Stern–Volmer quenching constant which turns out to be proportional to the lifetime of the fluorescent species in the absence of the quencher, τ_o.

For practical purposes we can replace the fluorescence efficiencies by the corresponding intensities I_f^o and I_f since $\phi_f^o = I_f^o/I_a$ and $\phi_f = I_f/I_a$. (The intensity of the radiation absorbed is independent of the concentration of the quencher).

If, therefore, we re-plot the data in Fig. 4.4a as $I_f^o/I_f - 1$ against the concentration we should get straight lines of slope K with a different value of K for each halide, increasing from Cl^- to I^-.

You can try this out shortly in a SAQ, but for the moment we will take a few measurements from the Cl^- curve given in Fig. 4.4a. Check the following data – you will be using it later.

$[Cl^-]$/mol dm^{-3}	Fluorescence Intensity
0.00	100
0.01	30.05
0.02	20.7
0.04	11.0
0.06	6.7
0.08	5.4

Π What would be the effect on the analysis of changing the fluorescent species to one of longer life-time?

It would increase the sensitivity of the analysis since the slope of the calibration graph, K, is proportional to the lifetime of the excited state. This is what we might expect from basic principles since, as we have already seen in Part 1, the longer the lifetime of the excited state the greater the chance of it being deactivated by collision with a quenching species. Consequently the same degree of quenching can be achieved with a lower concentration of quencher when a longer time is available.

The most important application of quenching is in the determination of oxygen which, as we have already seen, is well-known for its properties as a quencher – though usually for its nuisance value! There are many operations which have to be carried out in the absence of oxygen and a supply of oxygen-free nitrogen or inert gas is required to provide a suitable atmosphere in a glove box or similar environment. Fluorescence quenching provides a possible method for monitoring low levels of oxygen in such a situation.

For this application a slightly modified version of the Stern–Volmer equation is used which relates the fluorescence measurement directly to the partial pressure of oxygen in the atmosphere in equilibrium with the sample. The equilibrium is set up by bubbling the gas through the solution of the fluorescer in a silica cell. As before the sensitivity of the method is proportional to the lifetime of the fluorescent species, τ_0. With pyrene ($\tau_0 \approx 10^{-6}$ s), the fluorescence intensity is reduced to half the original value ('50% quenching') by a partial pressure of oxygen of about 1%. With eosin ($\tau_0 \approx 10^{-3}$s), 50% quenching is achieved with about 10 ppm of oxygen. Thus with a suitable choice of fluorescent species a wide range of oxygen concentration can be covered.

The long lifetimes of the emission from pyrene and eosin arise from processes which involve inter-system crossing to the triplet state followed by conversion back to the singlet state from which the emission actually originates. This phenomenon is referred to as 'delayed fluorescence'.

Π Another indirect fluorimetric method often referred to (incorrectly) as a quenching method is one which the ligands in a fluorescent complex are replaced by an analyte species to give a complex which is not fluorescent. A well-known example of this type is the determination of fluoride ions which replace the organic ligand in a number of aluminium complexes such as those with oxine, morin, and alizarin garnet. As with true quenching methods the fluorescence intensity falls as the concentration of fluoride increases. Why is it incorrect to call this a 'quenching method'?

The fluorescence decreases because the fluorescent species is actually being destroyed by a chemical reaction with the analyte. In a true quenching method the fluorescent species is deactivated and it is present at fixed concentration throughout.

The reverse process, in which an analyte reacts with a non-fluorescent complex releasing the reagent which then forms a fluorescent complex with another metal, has also been used as an indirect method for CN^-. The cyanide ion replaces oxine in the non-fluorescent palladium-oxine complex which then forms a fluorescent complex with magnesium ions. This method is however subject to interference from other ions such as sulphides and thiocyanates which do not affect the direct method with benzoquinone described earlier.

SAQ 4.4a

The fluorescence of quinine is quenched by the presence of chloride ions, and the following data obtained.

⟶

SAQ 4.4a (cont.)

$[Cl^-]$/mol dm^{-3}	Fluorescence Intensity, I_f
0.00	100 $[I_f = I_f^o]$
0.01	30.05
0.02	20.7
0.03	14.7
0.04	11.0
0.05	8.2
0.06	6.7
0.07	5.9
0.08	5.4

Plot the data using the Stern–Volmer equation. Do you get a straight line?

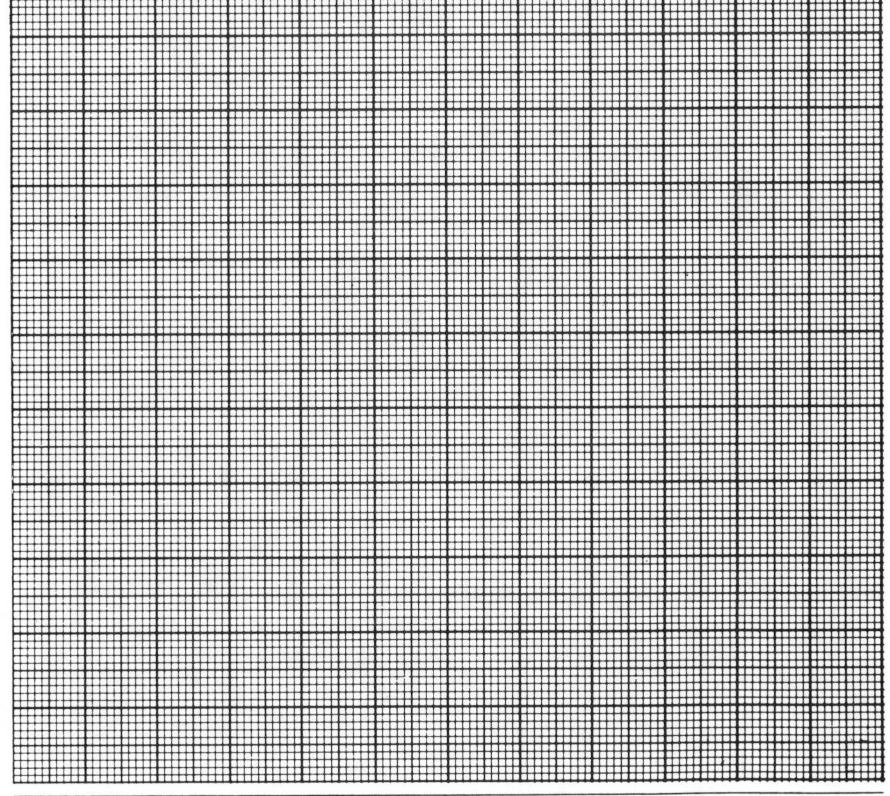

SAQ 4.4b

Select the phrase, (*i*) to (*iv*), which correctly completes the following sentence:

Chloride ions cannot be determined in the presence of bromide ions by the quenching of quinine fluorescence because

(*i*) the lifetime of the excited quinine molecule is too short.

(*ii*) the sensitivity increases with the mass of the ion.

(*iii*) quenching techniques are not specific.

(*iv*) the calibration graph is non-linear.

SAQ 4.4c

Select the correct answer:

To obtain 50% quenching with 0.1% of oxygen would require a fluorescer having a fluorescence lifetime of

(*i*) 10 μs
(*ii*) 0.1 μs
(*iii*) 10 ms
(*iv*) 0.1 ms

4.5. SCANNING OF FLUORESCENCE SPECTRA

In conventional fluorescence analysis the majority of quantitative measurements are made using fixed wavelengths for excitation and emission. As we saw in Part 2 it is also possible to record excitation and emission spectra with instruments having automatic scanning monochromators.

Π Explain briefly how you would record the emission spectra of a sample using a dual monochromator spectrophotometer

Set the excitation monochromator to the wavelength of maximum excitation and scan the emission monochromator from as close to that wavelength as possible to longer wavelength until the emission intensity falls to the baseline value.

4.5.1. Quantitative Analysis

Π How would you use the fluorescence emission spectrum in quantitive analysis? What advantage does this have over the fixed wavelength procedure and what disadvantage?

Measure the height of the emission peak of the sample above the baseline and convert this to a concentration value with a calibration graph prepared using similar measurements made on a set of standard solutions. The advantage is that any variation in background emission from one sample to another can be seen and a correction made. The presence of another fluorescent species in varying amounts can also be seen and allowance made for it. Indeed random variations of this type in a particular set of samples would make the fixed wavelength procedure very inaccurate. The disadvantage, of course, is that this type of measurement takes much longer. In general analytical practice, the full excitation or emission spectra are used during the development of an analytical method in order to detect any potential problems but, once the method has been shown to be reliable for the purpose for which it is intended, routine measurements are carried out with fixed wavelengths – often as part of an automatic procedure handling perhaps 100 samples per hour where the analytical demands warrant it.

4.5.2. Qualitative Analysis

The other situation in which it is necessary to record a full spectrum is in qualitative analysis. Unfortunately, fluorescence spectra, like uv absorption spectra, are not very characteristic of the compound from which they are obtained and in general it is not possible to identify a completely 'unknown' compound from either the excitation or the emission spectra. For this purpose infrared and nmr spectra are much more successful.

Π Why do both fluorescence and uv absorption spectroscopy suffer from this limitation in qualitative analysis?

Both techniques depend upon transitions between electronic energy levels. The actual energy values are determined by the π-system and so the bands observed for all compounds containing the same basic π-system (C=C, C=C—C=O, the benzenoid ring etc.) occur in the same narrow region of the spectrum. The presence of different groups in the molecule remote from the π-system does not affect the absorption properties and even directly attached substituents have little effect unless they are polar and affect the electron distribution significantly. Even then the effect is small and results only in a change in intensity or a small shift in wavelength. Electronic spectra are also very simple consisting of one or two broad bands having a smooth contour in most cases. The success of infrared spectroscopy on the other hand arises from the very much wider spectral range and the sharpness of the absorption bands which are due to vibrational transitions and are often characteristic of the presence of particular functional groups.

The presence of vibrational structure in an electronic band will also enhance the capabilities of electronic spectroscopy for qualitative analysis because it provides us with far more experimental data – several different peak maxima and intensities instead of only one for a smooth peak with no fine structure. Consequently the spectrum is far more characteristic of the compound from which it was obtained and thus makes it possible to recognise that compound when it occurs as an 'unknown'.

Π Name a compound for which the electronic bands show vibrational fine structure.

The most common examples are aromatic hydrocarbons, particularly when several aromatic rings are fused together to form a large planar chromophore. We have frequently referred to anthracene as a typical fluorescent molecule and you should be familiar with the pronounced fine structure in both its emission and excitation spectra.

Π Unfortunately, the existence of fine structure in fluorescence spectra is relatively uncommon. When it occurs with fluorescent derivatives rather than as the natural fluorescence of the analyte it may not be sufficiently characteristic to identify the unknown. Why should this be?

With derivatives it is the derivatising molecule that provides the chromophore and so unless different analytes affect the electron distribution to a different extent the electronic spectra will be more characteristic of the derivatising molecule. This is particularly true of reagents used as fluorescent labels – a property which, as we shall see, is highly desirable in the use of fluorescence in hplc detection.

4.5.3. Dual Wavelength Spectroscopy (Synchronous Scanning)

Π If a uv absorption spectrum is to be used in qualitative analysis, the sample is scanned across the whole range of the spectrophotometer (200–450 nm or to higher wavelength if the sample is coloured) and bands will be observed corresponding to any absorbing species present. Unfortunately a problem arises in fluorescence spectroscopy which prevents us using conventional excitation and emission spectra in this way. What is it?

Fluorescence involves controlling two wavelengths independently. In order to observe emission spectra we must excite at the correct wavelength for a particular compound. If the compound is an 'unknown' we should not know what excitation wavelength to choose nor the wavelength at which to observe the emission. It would therefore be necessary to carry out a preliminary test run using the 'direct light' setting of the emission monochromator in order to find the best wavelength for excitation. With a mixture of fluorescent species, this procedure would identify several wavelengths for excitation and it would then be necessary to run emission spectra at each of these excitation wavelengths. This would be a lengthy process though, in the end, it might prove more successful than simple absorption spectroscopy since a different response is frequently break obtained for compounds whose absorption spectra overlap seriously. Without the direct light facility (on an old prism instrument for example) it would not be possible to determine excitation wavelengths in this way and the only safe procedure would be to scan the emission spectra repeatedly with a small increase of excitation wavelength each time. This would be an even more lengthy process.

An alternative approach is to scan both monochromators simultaneously with a fixed wavelength between them. For many compounds the Stokes' shift is around 50 nm and so if the monochromators are scanned at the same speed with a difference of 50 nm between their settings the emission monochromator should pick up some of the emission from each compound as the excitation monochromator passes through its excitation band.

Π What is meant by the Stokes' shift? What would the emission monochromator be reading in the example just described when the excitation monochromator was at 350 nm?

The Stokes' shift is the wavelength difference between the excitation maximum and the emission maximum. The emission monochromator would be at 400 nm, emission always being at a longer wavelength than excitation. The spectrum which is obtained by this procedure, known as 'synchronous scanning', is quite different from either the excitation or emission spectrum though it is obviously derived from them. Furthermore its appearance will change if we alter the wavelength difference between the two monochromators. Let's take a simple example and see what happens.

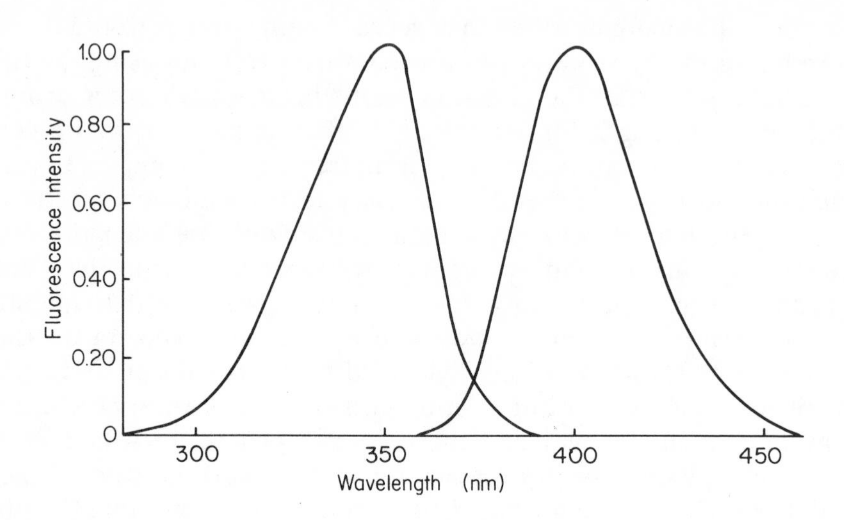

Fig. 4.5a. *The excitation and emission spectrum of a fluorescent compound having its excitation maximum at 350 nm and emission maximum at 400 nm*

Remember that the excitation spectrum is a plot of the intensity of the emission at 400 nm as we vary the wavelength of excitation. The intensity scale is adjusted to give a value of 1.00 unit at the excitation maximum. If we move the excitation monochromator to 340 nm the intensity falls to 0.84. If we now move the emission monochromator to 390 nm the intensity falls to 0.84×0.77. So, with a synchronous scan with a 50 nm difference between the monochromators (which is exactly equal to the difference between the wavelengths of maximum excitation and emission) the intensity when the excitation wavelength is 350 nm is still 1.00. The value at an excitation wavelength of 340 nm, however, is only 0.65. Values at other wavelengths can be derived similarly by taking products between the intensities on the excitation and emission spectra at 50 nm intervals across the spectra as shown in Fig. 4.5b.

Excitation spectrum		Emission spectrum		Synchronous scan spectrum Intensity with spacing (λ_{em} − λ_{ex})			
λ_{ex}/nm	Int.	λ_{em}/nm	Int.	40nm	50nm	60nm	70nm
300	0.05	350	0.00	0.00	0.00	0.00	0.00
310	0.16	360	0.00	0.00	0.00	0.00	0.05
320	0.37	370	0.05	0.03	0.02	0.01	0.28
330	0.60	380	0.31	0.26	0.19	0.11	0.60
340	0.84	390	0.77	0.77	0.65	0.46	0.70
350	1.00	400	1.00	0.71	1.00	0.84	0.51
360	0.71	410	0.84	0.18	0.60	0.84	0.19
370	0.21	420	0.51	0.02	0.11	0.36	0.03
380	0.04	430	0.27	0.00	0.01	0.06	0.00
390	0.00	440	0.12	0.00	0.00	0.00	0.00
		450	0.03				

Fig. 4.5b. *Data for synchronous scan of a fluorescence spectrum*

When these values are plotted against the excitation wavelength we get curve A in Fig. 4.5c. Notice that the maximum is still at 350 nm and of intensity 1.00 but that it falls off much more rapidly on either side than the excitation band to give a much sharper profile. If we carry out similar calculations for 60 nm and 70 nm spacings between the monochromators the peak maximum gets progressively lower and shifts away from the excitation maximum in a somewhat irregular manner depending on the shape of the excitation and emission bands. This is shown in curves B and C in Fig. 4.5c.

Π What happens to the band-width as the wavelength interval increases?

It gets larger. However, over the small range we have covered the band-width is still less than that of the excitation band.

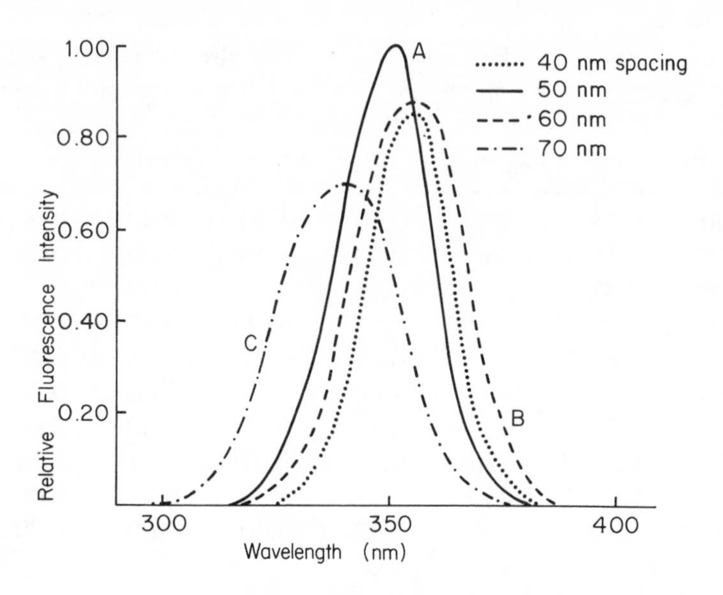

Fig. 4.5c. *Synchronous scan spectra recorded under four different conditions*

Synchronous scanning therefore avoids the problems of not knowing the wavelength at which the excitation monochromator should be set and also reduces the band-width of the peak obtained for a given compound, particularly if we happen to choose a wavelength difference close to the Stokes' shift. In this respect the effect is very like that obtained by using derivative spectroscopy in absorption work where the difference in absorption between two wavelengths of fixed interval is plotted instead of the simple absorption. Obviously if we have a mixture of analytes we cannot choose the optimum wavelength separation for each component simultaneously but, unless there is an unusual range in the Stokes' shifts, it should be possible to get some response from each component.

Although the 'spectrum' obtained by synchronous scanning has no fundamental significance and varies with the wavelength separation, it does provide a characteristic 'fingerprint' for a particular compound, particularly if its fluorescence spectra shows fine structure. If we create a reference library of such spectra for a series of compounds of interest, all run at the same wavelength separation and monochromator bandwidth, then we should be able to recognise any of them in mixtures of 'unknowns' and if necessary determine the concentration.

This type of analysis has been used with some success to identify commercial petroleum products. These contain aromatic hydrocarbons and the other fluorescent species in different proportions according to the source of the material. This complex mixture produces a synchronous scan spectrum which can be used to identify a particular batch of fuel oil for example or to prove that two samples were obtained from the same original bulk supply. This is of particular interest in criminal investigations and has now become an established technique in forensic laboratories.

We are now going to compare some 'normal' emission spectra with their associated synchronous scan spectra, in order to illustrate what synchronous scanning has to offer.

SAQ 4.5a

Fig. 4.5d shows the normal emission and the synchronous scan spectra of anthracene. What are the three chief differences between these two spectra?

⟶

SAQ 4.5a (cont.)

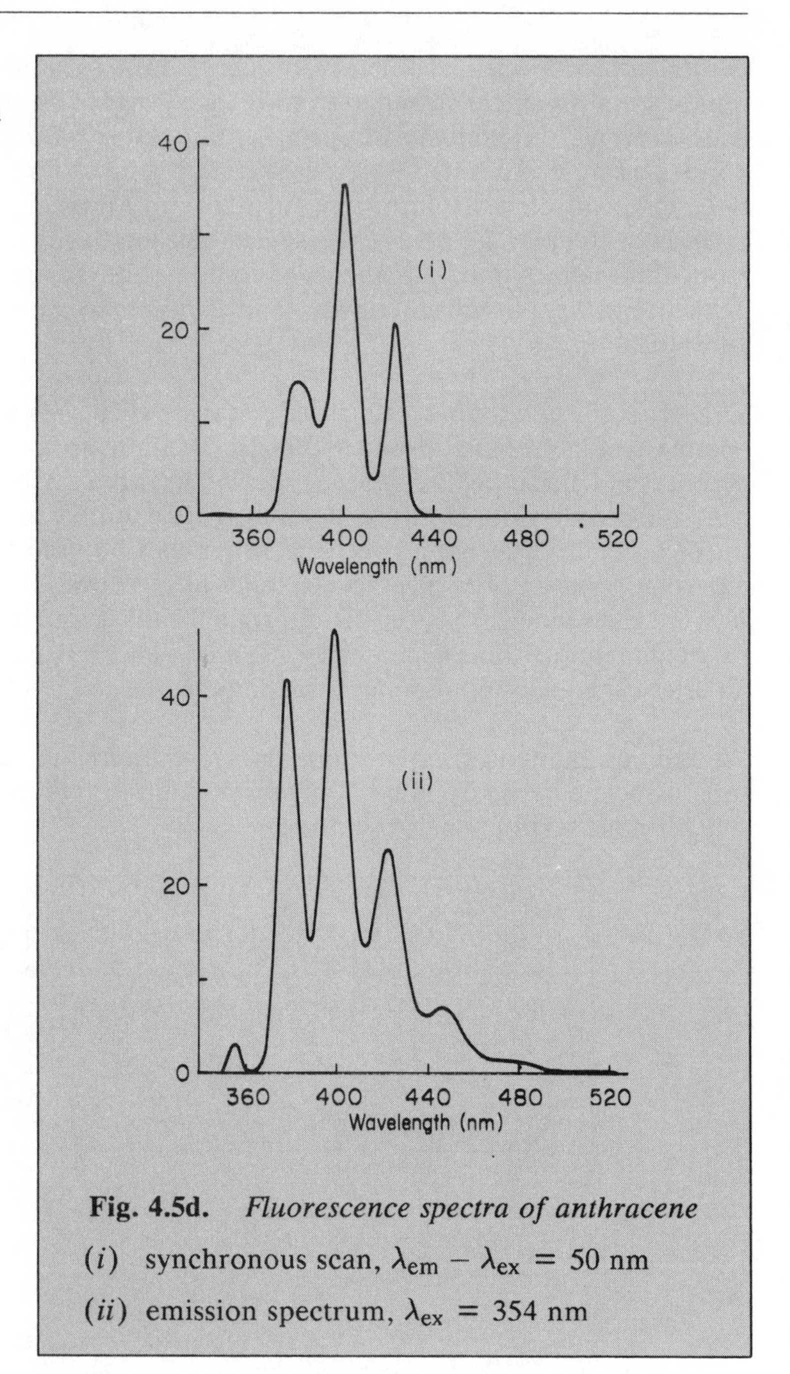

Fig. 4.5d. *Fluorescence spectra of anthracene*

(*i*) synchronous scan, $\lambda_{em} - \lambda_{ex} = 50$ nm

(*ii*) emission spectrum, $\lambda_{ex} = 354$ nm

SAQ 4.5a

SAQ 4.5b

What advantages do the following two differences between an emission spectrum and a synchronous scan spectrum confer on the synchronous scan spectrum?

(*i*) The band-width of a synchronous scan spectrum is less than that of an emission spectrum.

(*ii*) There is no Rayleigh peak in synchronous scan spectrum.

SAQ 4.5c

Fig. 4.5e shows fluorescence spectra of a mixture of fluorene, naphthalene and anthracene recorded under three different conditions. In what ways do these spectra demonstrate the advantages of the synchronous scanning technique when examining a mixture of fluorescent compounds?

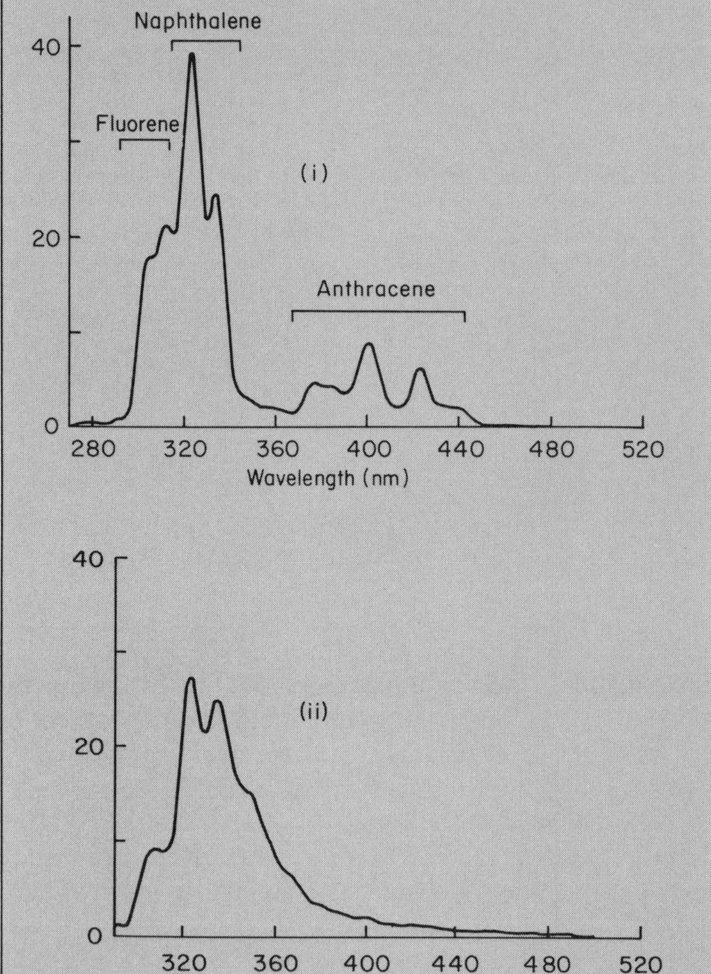

⟶

SAQ 4.5c (cont.)

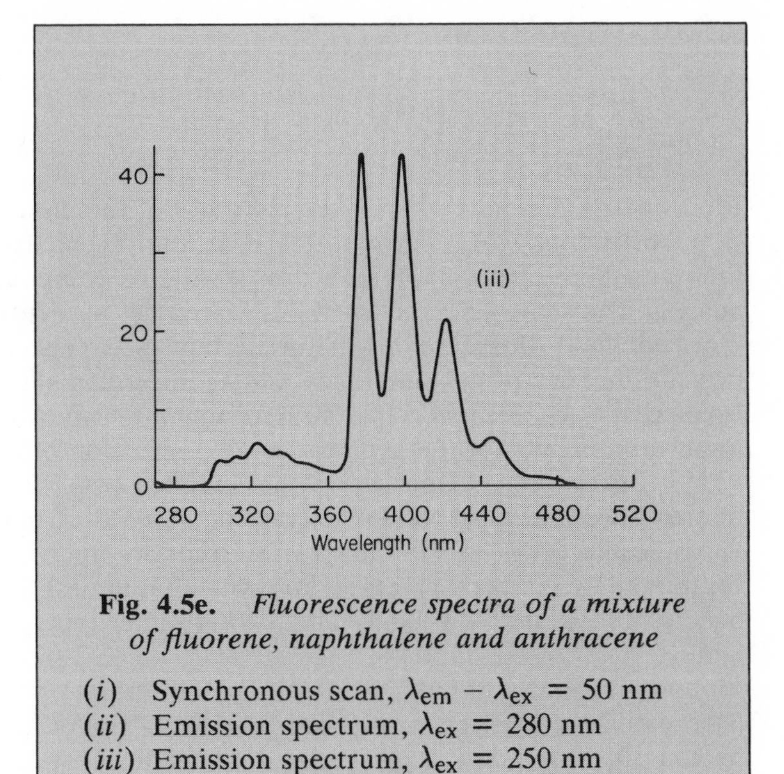

Fig. 4.5e. *Fluorescence spectra of a mixture of fluorene, naphthalene and anthracene*

(*i*) Synchronous scan, $\lambda_{em} - \lambda_{ex} = 50$ nm
(*ii*) Emission spectrum, $\lambda_{ex} = 280$ nm
(*iii*) Emission spectrum, $\lambda_{ex} = 250$ nm

SUMMARY AND OBJECTIVES

Summary

Fluorimetric methods of analysis make use of the natural fluorescence of the analyte, the formation of a fluorescent derivative or the quenching of the fluorescence of a suitable compound by the analyte. The scope of direct methods is limited, particularly in the inorganic field where the only naturally fluorescent species in solution are the ions of the lanthanide and actinide elements and a few isolated species. Emission is usually of low intensity and the most important analysis of this type is the determination of uranium as UO_2^{2+}. Many organic species can be analysed using their natural fluorescence and methods for polyaromatic hydrocarbons, several vitamins and drugs and a number of steroids are in common use in routine analysis. In other cases, however, the natural fluorescence occurs at relatively short wavelength where interference is a problem.

Fluorescent derivatives for metals in the main groups are obtained by complex formation using reagents similar to those employed in colorimetry. Reagents can be made more specific by the use of solvent extraction or by control of pH. They can also be designed so that steric factors allow them to react with only a limited range of elements. Complexes of the transition elements on the other hand are usually non-fluorescent.

Complexes of the lanthanide elements are of interest not so much because they provide a method of analysis for these elements but because they can be used in the determination of organic compounds with the lanthanide acting as fluorescent label. The emission from these complexes consists of sharp bands because the transitions originate in excited 'atomic' states. This emission is also long-lived which enables time resolution techniques to be used to eliminate scatter and reduce interferences in biological samples. This is becoming important in protein analysis.

Preparation of derivatives of non-metals and anions involve more varied chemistry. They are important because species of this type are not easily determined by atomic spectroscopy.

The formation of derivatives of organic analytes usually involves condensation or ring closure reactions. Different reagents are required for each type of compound so that the method is, by its very nature, specific.

Quenching methods in general are completely non-specific but often useful where samples are known to contain only one species able to quench fluorescence. They are occasionally used for halide ions but the most important application is the determination of oxygen. Since the analytical signal decreases non-linearly as the concentration of the analyte rises mathematical processing of the results using the Stern–Volmer equation is useful to obtain a linear calibration. The sensitivity of the method can be increased if the fluorescer has an excited state of relatively long lifetime.

The scanning of fluorescence spectra is advantageous when developing a quantitative procedure or for measurements on samples where background levels vary or where interfering species are likely to be present. It is also essential for qualitative work. Fluorescence spectra suffer from the same limitations as uv absorption spectra but can provide a positive means of identification when bands show vibrational fine structure.

The use of synchronous scanning, where both monochromators are scanned together at a fixed wavelength difference, is particularly useful for analysing mixtures where the excitation wavelengths are unknown and for providing a 'fingerprint' for commonly occurring compounds in commercial products.

Objectives

You should now be able to:

- distinguish between methods involving measurement of the natural fluorescence of the analyte, derivatisation and quenching;

- name the inorganic species which can be analysed by direct methods;
- recall the electronic structure of the lanthanides and actinides and explain the origin of their fluorescent properties;
- describe the chief types of organic compounds which can be analysed using their natural fluorescence;
- explain why, in general, natural fluorescence is only useful in analysis when it occurs at long wavelengths;
- recognise that the principle of complex formation used in colorimetry is equally applicable to fluorimetry and that advantages can be gained by using the latter technique;
- describe the principal differences in the fluorescent properties of complexes of metals in the main groups, the transition series and the lanthanide series;
- name ways of making fluorimetric procedures more selective;
- recognise the potential of the lanthanides as probes for the analysis of biological materials and explain the reason for this;
- describe the principles involved in forming fluorescent derivatives of organic compounds and quote some specific examples;
- explain the advantages of synchronous scanning in qualitative analysis;
- outline methods for the fluorimetric determination of a wide range of analyte species, including non-metals, anionic species, molecular oxygen and polyaromatic hydrocarbons.

5. Further Applications of Photoluminescence

5.1. INTRODUCTION

In Part 4 we looked at the applications of photoluminescence to analysis in general terms and you should now be familiar with the potential of the technique and the types of analytical problem for which it might be used in the real world. In Part 5 we shall extend this survey to some more specialised techniques, particularly those which are likely to become of increasing importance in the near future as instrumentation develops and the requirements for analysis become more demanding.

We shall also examine in detail some specific examples of photoluminescence methods taken from the scientific literature to give you a better appreciation of the more practical aspects of fluorescence measurements in analysis. These 'case studies' are chosen to illustrate different types of analysis: organic, inorganic, fluoroimmunoassay (FIA) and room temperature phosphorescence (RTP). This will also provide an opportunity to revise material presented earlier in the Unit and general analytical principles. The associated SAQs will give you direct experience of working through a fluorimetric analysis calculation in detail.

Although the case studies are placed together at the end of Part 5 for convenience, you should not attempt to work through them all in succession. The first two, Sections 5.5 and 5.6 do not rely on the earlier sections and you could well tackle one of these before going on to the new material in Section 5.2. Similarly, you will probably find it advantageous to work through the case studies on FIA and RTP immediately after reading the sections describing the principles of these techniques (ie Section 5.7 after 5.3 and Section 5.8 after 5.4).

5.2. FLUORESCENCE AND CHROMATOGRAPHY

5.2.1. Plane Chromatography

Fluorescence has been used for many years as a means of locating the components of a mixture separated by paper or thin layer chromatography (tlc). There are three ways of doing this. All of them are applications of the methods we have discussed earlier though two of them are perhaps more obvious than the third.

Π Describe the processes by which the components could be made visible using fluorescence (or, at least, the two more obvious ones!)

In each case the chromatogram will be viewed under an ultraviolet lamp in conditions of subdued lighting.

(*a*) If the compounds separated by the chromatographic method are naturally fluorescent with emission in the visible region they will be clearly seen as coloured 'spots' without any further treatment.

(*b*) If the compounds are not fluorescent or emit only in the uv region it will be necessary to spray the chromatogram with a solution containing a suitable reagent to form a fluorescent derivative with them. If the sample contains compounds of widely differing chemical constitution, it may be necessary to spray with two or more different reagents to derivatise all the components present. This method is, of course, completely analogous to the more usual method of visualising spots in plane chromatography by spraying with reagents to form coloured derivatives. Like the colorimetric method, the colour of the fluorescence emission often gives a useful indication of the identity of the compounds present. (These two methods are more satisfactory with thin layer chromatography than with paper chromatography because paper often gives a significant fluorescent emission under uv radiation. The materials used as the stationary phase in tlc tend to give much lower background emission.)

(*c*) If the paper or thin layer plate has significant background fluorescence and the compounds present are not fluorescent in the visible region, they will show up as dark spots on the chromatogram where the background fluorescence appears to be quenched. In most cases this is not true quenching but an extreme inner-filter effect. The molecules of the compound present in the spot absorb the incident radiation before it can excite the fluorescence of the background material. This has become the most important method of visualising spots in thin layer chromatography and many laboratories now purchase thin layer plates containing a fluorescent additive such as manganese-activated zinc silicate which absorbs at 254 nm and emits strongly in the visible region. Since many compounds absorb uv radiation at this wavelength, the method has a wide range of application.

Π Why is this last visualisation technique, which is totally non-specific, so successful? (There are two main factors.)

We have already mentioned its wide range of application with so many compounds absorbing at 254 nm. The other factor is that, because the technique is non-specific, it is particularly suitable for chromatographic work where the components of the mixture have already been separated from each other. The chromatography provides the selectivity and a uniform response from the detection system (visualisation) is a decided advantage.

The identification of spots can be made much more positive by analysing the emission with a spectrofluorimeter. This can be done with a special sampling accessory using frontal illumination. The tlc plate is first scanned by driving it systematically across the instrument sampling position in order to locate the position of the spots.

Π What would be the most effective settings for the two monochromators during this operation?

Excitation monochromator – the choice of a wide band-width and a wavelength about 230 nm will provide the most favourable conditions for exciting the fluorescence. Emission monochromator – the direct light position should be used (if available) with a cut-off filter to exclude scattered radiation at the exciting wavelength. This will make it possible to use a fairly wide band-width to enhance the sensitivity. Otherwise, the best option is to set the wavelength at 300 nm and use as wide a band-width as possible. Once the spots have been located they can be examined one at a time with narrow band-widths and the most favourable wavelengths determined for recording the emission and excitation spectra. A positive identification of the compounds may then be possible (allowing for the shortcomings of electronic spectra in this respect). Since the compounds are in the solid phase the possibilities of identification can be increased by the observation of phosphorescence where the instrumental facilities allow for it.

5.2.2. Column Chromatography

The development of high performance liquid chromatography (hplc) has been perhaps the most important advance in analytical chemistry in the past ten years. One of the most crucial features in this technique is the detection of components as they are eluted from the column. Since the sample size is commonly 1–10 μl of a solution containing perhaps 1–10% of a mixture of many constituents, the detector system has to be capable of responding to less than
1 μg of material. The success of hplc is therefore critically dependent on the efficiency of the detector.

Ultraviolet detectors have been very widely used because so many compounds absorb radiation or can be readily converted into strongly absorbing derivatives.

Π What advantage would there be in using a fluorescence detector?

A gain in sensitivity of two or three orders of magnitude. Fluorescence detectors have now become well-established in hplc work even though they are not so widely applicable as uv detectors. Both direct and derivatisation methods are used.

Π What properties are desirable in a derivatising agent for hplc?

The derivatives should all have similar fluorescence characteristics. As with tlc, the chromatography provides the selectivity so a uniform response to each separated component would be an advantage. This occurs when a mixture of compounds of a particular class are derivatised with the same fluorescent derivatising agent. Some examples are given in Fig. 5.2a.

Modern microprocessor-controlled liquid chromatographs can be programmed to change detector wavelengths after pre-set time periods corresponding with the retention times of particular components. This extends the range of compounds that can be detected during a single run when there are wide variations in their fluorescence characteristics.

Where derivatisation is used, there are two possible lines of approach. The simplest is to form the derivatives of all the species present (or as many as are of interest) by adding the reagent to the sample before it is injected on to the column. This is particularly useful where the chemistry of the derivatisation reaction is at all complicated requiring, perhaps, the addition of several reagents, heat treatment and time for completion of the reaction. This is called 'pre-column derivatisation' and is the preferred technique though it can give rise to problems by modifying the chemical properties of the analytes so that the separation is less satisfactory.

Where this problem is acute, it is necessary to turn to 'post-column derivatisation' in which the derivatising agent is injected continuously into the eluate from the column.

Π What problems might arise with post-column derivatisation which do not affect pre-column derivatisation?

(*a*) The reagent is flowing continuously through the detector. This may raise the background fluorescence and reduce the sensitivity since the reagent will have to be at a much higher concentration than the analyte to ensure efficient conversion to the derivative. It might also cause excessive absorption in the fluorimeter beam and distortion of the fluorescence signal.

(*b*) It may be necessary to insert heating coils or delay coils to ensure that the reaction is complete before the components reach the detector. This additional path-length between column and detector will allow the components to diffuse and spread

Reagent	*Analyte*
4-bromomethyl-7-methoxycoumarin	carboxylic acids
7-chloro-4-nitrobenzyl-2-oxa-1,3 diazole	amines and thiols
dansyl chloride	primary amines and phenols
dansyl hydrazine	carbonyl compounds

dansyl =
1-dimethylaminonaphthalene-5-sulphonyl

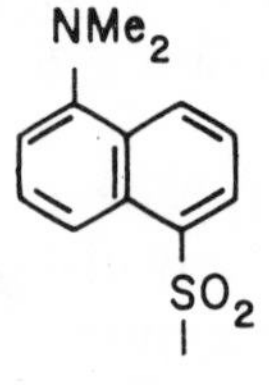

NMe_2

SO_2

Fig. 5.2a. *Reagents used in chromatography with fluorescence detection*

out in the eluate and impair the resolution of the separation. Wherever possible, therefore, pre-column derivatisation is the method of choice though the range of suitable reagents is limited.

All the reagents employed in standard fluorimetric procedures can be adapted for hplc work. However, since their effect on the separation can be critical, several new reagents have been developed specifically for this purpose. Some examples are given in Fig. 5.2a.

SAQ 5.2a

Why is the lack of specificity in a derivatising agent

(*i*) no problem in either tlc or hplc,
(*ii*) often an advantage in hplc?

SAQ 5.2b

When using a tlc plate scanning accessory to locate fluorescent spots on a tlc plate, which of the following settings would you use for:

(*i*) the excitation monochromator;
(*ii*) the emission monochromator?

— wide slit-width
— narrow slit-width
— short wavelength
— long wavelength
— direct light setting
— cut-off filter to exclude short wavelength

SAQ 5.2c

State whether each of the following effects resulting from the formation of a fluorescent derivative is:

(*A*) associated with pre- or post-column derivatisation;
(*B*) an advantage or a disadvantage in an analysis.

(*i*) Absorption of the eluate is increased.
(*ii*) Detection of the eluted components is delayed.
(*iii*) The efficiency of the separation is affected.
(*iv*) All the analytes are derivatised simultaneously.
(*v*) Conditions for the optimum separation of the analytes are not affected.

SAQ 5.2d

Draw the structure of the dansyl chloride molecule. Why is it a satisfactory derivatising agent for fluorescence detection in hplc?

5.3. FLUOROIMMUNOASSAY

Another type of derivatisation, of increasing importance in analytical chemistry in the biological field in particular, is the antibody/antigen reaction upon which we (and the rest of the animal kingdom) depend to protect us against infection from disease. The antibodies responsible for the effective functioning of this 'immune system' are typically very large protein molecules which circulate in the blood-stream and remove the threat posed by 'antigens', the general name given to bacteria, viruses and other 'foreign' bodies, by forming a complex with them. This is then safely eliminated from the system.

Π The most remarkable feature of the antibody/antigen reaction is its very high specificity and it is this property which makes the technique, referred to as 'immunoassay', so attractive to analysts. It provides us with a highly specialised 'reagent' which will react with only one component in a complex mixture and gives us the opportunity of carrying out a quantitative analysis for that compound virtually free from interference from the other species present. There are, however, two major problems to be overcome first. Try to predict what these might be.

(*a*) We have first of all to produce the antibody to react with the analyte we are trying to determine. The analyte will be the antigen in the system.

(*b*) We have to devise some means of determining the extent to which the reaction between the antibody and the antigen has proceeded.

Let's deal with the second, analytical, point first as it more familiar ground.

5.3.1. The Antibody/Antigen Reaction

We may represent the 'immune reaction' by the equation

Ab	+	Ag	=	(Ab–Ag)
antibody	+	antigen	=	complex

The antibody–antigen complex usually precipitates out from the reaction medium and analytical methods involving separation have been widely used. The use of labelled species has been found to have many advantages, in particular, making it possible to follow the reaction using instrumental methods which can be readily automated. However, most instrumental methods measure the molar concentration of the analyte and this can be a problem with species of high relative molar mass.

Π Suppose we had a solution containing 1.0 g dm^{-3} of an immunoglobulin of M_r 150 000 which is a typical antibody found at this level in human serum. What is its molar concentration?

0.000 0066 mol dm^{-3}. Since 1.0 mol dm^{-3} is 150 000 g dm^{-3}, 1.0 g dm^{-3} is 1/150 000 g dm^{-3}.

Clearly we are going to need a very sensitive technique to monitor concentrations below 10^{-5} mol dm^{-3} which would be pushing uv detection, for instance, to its limit. The high sensitivity associated with fluorescence measurement makes it a promising candidate but, although it was one of the first techniques to be tried in this application, it was radiochemical techniques with their even more spectacular sensitivity which eventually became established as the more successful approach.

The usual procedure adopted with this technique, which is referred to as 'Radioimmunoassay' (RIA), is to follow the competition between the antigen, Ag (the analyte), and a known weight of a radioactively labelled version of the same antigen, Ag*, for a limited amount of the appropriate antibody, Ab (ie the total antigen is in excess). The antigens form a complex with the antibody which is insoluble in the reaction medium and has to be separated by centrifuging or filtering. This step is often difficult and a number of special techniques have been developed to assist in the separation. After separation, the distribution of radioactivity between the complex and the antigen remaining in solution is measured and the ratio used to determine the concentration of the inactive antigen. The reaction is represented by the equation

$$2Ag + 2Ag^* + 2Ab = \underbrace{(Ab\text{–}Ag + Ab\text{–}Ag^*)}_{\text{'bound'}} + \underbrace{Ag + Ag^*}_{\text{'free' or 'unbound' excess antigen}}$$

Π Assuming that Ag and Ag* compete equally for the antibody, how will the ratio of 'bound' to 'unbound' activity change as the concentration of inactive Ag originally present increases? (Remember that the amounts of Ab and Ag* are fixed.)

It will decrease. Since the amount of Ab is fixed and Ag is in excess, the amount of the complex (Ab–Ag) is also constant. When there is no inactive Ag present all the complex is in the active form, Ab–Ag*. As the concentration of inactive Ag increases it replaces some of the active Ag* in the complex and so the activity of the complex decreases while the activity of the solution rises. Hence the ratio of bound to unbound activity falls. The concentration of the analyte is readily calculated from algebraic relationships derived from the equilibrium equations but we do not need to go into this in detail.

Because the activity of radioisotopes can be measured at very low concentrations, the method has the potential for very low limits of detection. Indeed, it is regularly used to measure insulin and oxytocin in human blood where the levels normally present are 10^{-10} and 10^{-12} mol dm^{-3} respectively. However, there are some disadvantages – principally in terms of the handling of radioisotopes (though these are slight at the levels of activity used in RIA) and the associated cost and inconvenience of providing a safe environment in which to use them. The cost of counting equipment is also significant and many of the labelled compounds required are quite expensive. These labelled compounds often contain isotopes of relatively short half-life such as ^{125}I (60 days) so the shelf-life is limited. These disadvantages, together with the improved performance of modern fluorescence instruments account for the the growing interest in fluorescence immunoassay (FIA) where the reaction is followed by means of an antigen with a fluorescent label.

Π A further advantage would be realised in a situation where the fluorescence characteristics of the free labelled antigen were different from those of the complex. What is it ?

There would be no need to separate the free antigen from the complex. A determination of this type is called a 'homogeneous assay' and is much more rapid and convenient than the 'heterogeneous assay' with its separation stage as used in RIA.

5.3.2 Fluorescent Labels

The materials used as fluorescent labels need to have high fluorescence efficiencies and to absorb and emit at long wavelengths (in the visible) to avoid interference from the natural fluorescence present in the biological samples at shorter wavelengths. The scattering from samples of this type is also very considerable because of the presence of the compounds of high relative molar mass which are often colloidal. A wide separation between the exciting and emission

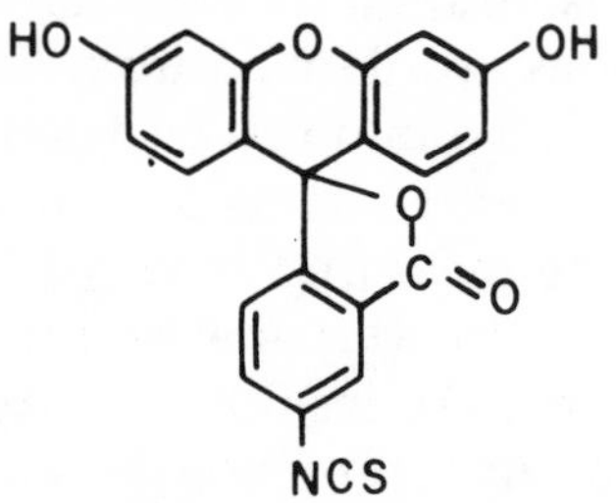

Fluorescein isothiocyanate (FITC)

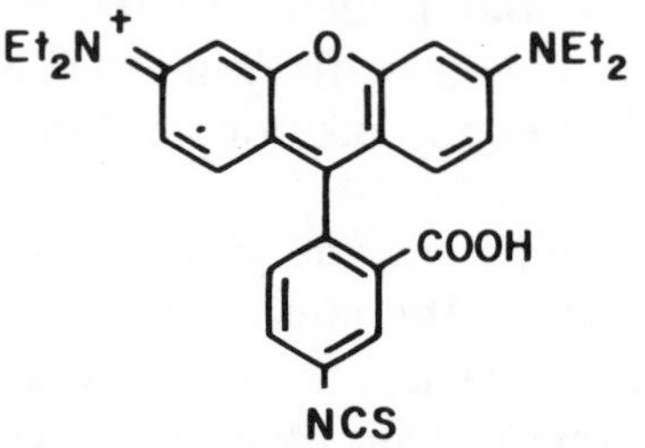

Rhodamine-B isothiocyanate (RBITC)

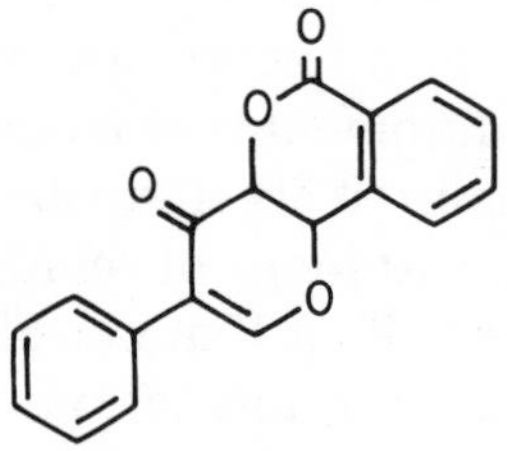

Fluoram

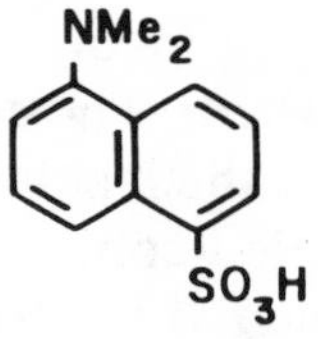

5-Dimethylaminonaphthalene-1-sulphonic acid (DNS)

Label	λ_{ex}/nm	λ_{em}/nm	ϕ_f	ϵ/dm^3 mol^{-1} cm^{-1}	Decay Time/ns
FITC	492	518	0.68	72 000	4.5
RBITC	550	585	0.30	12 300	3.0
DNS	340	500	0.30	3 400	14.0
Fluoram	394	475	0.10	6 300	7.0

Fig. 5.3a. *Commonly used fluorescent labels for proteins*

wavelengths (the 'Stokes' shift') is therefore another desirable feature. Some of the more common fluorescent labels for proteins are shown together with their fluorescent properties in Fig. 5.3a.

Π What new instrumental fluorescence technique which we have already discussed should be particularly beneficial as a means of discriminating against high background scatter and fluorescence?

Time resolution. For this purpose it is necessary to find a fluorescent label with a long lifetime – none of those in Fig. 5.3a would do. Our old friends the lanthanide chelates (eg europium salicylate) with their fluorescence lifetime of the order of 1 ms are particularly promising and it is in this area that they are most likely to realise their full potential. The fact that they have very large Stokes' shifts (about 250 nm), makes them useful for prompt fluorescence measurements too when time resolution is not available. Other complexes with lifetimes of around 1 μs (eg pyrene butyrate) can also be used but this is rather more difficult experimentally.

Π Apart from the spectroscopic properties, it is of course vital that the addition of the label should not alter the bonding properties between the antibody and antigen. Why would you think that this is more of a problem in FIA than in RIA?

Replacing an atom in the antigen molecule with its radioisotope does not alter its chemical or physical properties. A fluorescent label, however, is often a sizeable molecule in its own right and so adding it to the antigen molecule is quite likely to affect the properties of the antigen, particularly if it is close to the binding sites.

5.3.3. The Analytical Measurement

The chemical principles involved in FIA are identical with those in RIA and, though it is possible with both techniques to introduce variations in the procedure, we shall confine our attention to the example quoted earlier. With FIA, however, there is much more variety in the way in which the analytical measurement is made.

Π Even where the ratio of the signals from the complex and the solution is used in a heterogeneous assay it will not necessarily vary with the analyte concentration in the same way as the activity ratio in RIA. Why should this be?

The fluorescence efficiency of the complex containing the labelled antigen will not necessarily be the same as that of the free labelled antigen. Depending on circumstances, complex formation may increase or decrease the fluorescence efficiency or leave it unaltered.

Π What advantage can be taken of a change in fluorescence efficiency on formation of the Ab–Ag complex?

It makes homogeneous FIA possible because the total fluorescence intensity from the sample will vary as the ratio of bound to free Ag* varies. However, it will be necessary to plot a calibration graph of fluorescence intensity vs concentration of Ag using standards because algebraic treatment of the results becomes more complicated when different values for the fluorescence efficiencies have to be taken into account and uncertainties over their values may impair the accuracy of the method. There may also be problems with the measurement of prompt fluorescence from solid particles in suspension.

If the fluorescence efficiency of the complex is greater than that of the free labelled antigen, the technique is known as *fluorescence enhancement* when it is used in a homogeneous assay. If the fluorescence efficiency is less, the technique is referred to as quenching.

There are several alternatives to the simple measurement of fluorescence intensity which are used in fluoroimmunoassay. These include polarization measurements, energy transfer techniques and the use of enzymes to hydrolyse complexes and release fluorescent species. There are also methods involving chemi- and bioluminescence and in many cases somewhat complicated biochemical procedures are required. A full discussion of these methods is outside the scope of this Unit and you will have to follow up the references given in the bibliography if you want to learn the whole story!

Later on, in Section 5.7, you will be able to follow through a typical fluoroimmunoassay in some detail which should help to clarify the principles involved.

5.3.4. Preparation of the Antibody

Returning now to the other major factor in setting up an FIA method, we must consider how a suitable antibody can be produced to react with a particular analyte. The only practicable method is to make use of the natural processes occuring in the immune system of a suitable laboratory animal to produce the antibody. The principles of 'immunology', which is the name given to the relevant scientific discipline, are well-known and have been extensively developed, principally to produce vaccines against a wide range of diseases. You may well have read about this in the press (the subject is not without its controversial aspects!) and almost certainly you will, very early in your life, have received the benefits of this work with a vaccination against polio, whooping cough and measles.

The immune system produces antibodies as a response to the presence of the antigen in the body. When you are injected with polio vaccine, your body responds to the presence of the polio organism (antigen) by selecting a particular immunoglobulin molecule from the hundreds of thousands of different immunoglobulin is the antibody to the polio virus. It has binding sites which allow it to fit together exactly with the antigen molecule together exactly like two pieces of a jig-saw puzzle and this locks up the antigen out of harm's way as a complex. It is this exact match of the two structures which makes the antibody/antigen reaction so specific. In some cases, notably influenza and the common cold, there are many different viruses responsible for the disease and though the differences are very slight it is impossible to produce a vaccine which is effective against all of them.

Π How do you think this procedure is utilised to produce the antibody for a given fluorimunnoassay?

By injecting the analyte into a suitable laboratory animal (sheep and rabbits are commonly used). After a suitable time, blood is taken from the animal and the serum separated by conventional techniques. This is known as the 'antiserum' for the analyte and the process by which it is produced 'raising an antiserum'. Further separation using chromatographic technique is carried out to produce an antiserum in the required state of purity. In many cases, particularly when the analyte is a drug such as morphine, the analyte molecule is too small to induce an immune response in the animal – in which case it is referred to as a 'hapten'. It is then necessary to link it to a protein molecule such as albumin in order to raise the antiserum. A specific antiserum is referred to as anti-morphine, anti-thyroxine, etc indicating that it contains the antibody for the specified antigen. Confusion can sometimes arise when an immunoassay is used to determine levels of a particular human antibody such as immunoglobulin M (IgM). This is one of the commonest immunoassays used in hospital laboratories because it provides a means of following the progress of patients with particular diseases though at present it is usually carried out by RIA rather than FIA). In this case, the human antibody, IgM, is the antigen in the animal environment and the animal produces anti-IgM which becomes the antigen when added to human blood serum in order to determine the natural antibody already present. The word 'antiserum' is therefore useful to avoid the ambiguity in the other terms.

In practice the highly specialised methods necessary to raise an antiserum are not available to the analyst and a number of manufacturers now produce kits for specific immunoassays. These kits provide not only the antibody but also the labelled antigen (radioactive or fluorescent) and any other special solutions (buffers etc) required. Each kit will allow a stated number of assays to be carried out and, though expensive, they are cost-effective when a large number of assays are required.

The cost-effectiveness is much improved with modern centrifugal analysers fitted with fluorescence attachments. These require very much smaller reagent volumes (typically 250 μl) than conventional fluorimeters.

One recent development is to incorporate the antibody into solid particles which facilitates the separation of the Ab–Ag complex. Further ease of separation is achieved by making the particles magnetisable so that sedimentation on to a magnet can be used.

SAQ 5.3a

What are the chief advantages and disadvantages of each of the four fluorescent labels given below?

Label	λ_{ex}/nm	λ_{em}/nm	ϕ_f	ϵ/dm^3 mol^{-1} cm^{-1}	Decay Time/ns
FITC	492	518	0.68	72 000	4.5
RBITC	550	585	0.30	12 300	3.0
DNS	340	500	0.30	3 400	14.0
Fluoram	394	475	0.10	6 300	7.0

SAQ 5.3b

Write the equation for the antibody (Ab) – antigen reaction in conditions where there is an excess of antigen, labelled (Ag*) and unlabelled (Ag), competing for sites on the antibody. What is the analyte in this system?

SAQ 5.3c

In what respects does FIA score over RIA? (Four points were noted in the text.)

SAQ 5.3c

SAQ 5.3d

Explain why homogeneous assay is possible with FIA but not with RIA.

SAQ 5.3e

What reagents and solutions would you expect to find in a fluoroimmunoassay kit?

SAQ 5.3f

Are the following statements true or false?

(*i*) The chief attraction of the immunoassay technique is its high sensitivity.

(*ii*) Labelling techniques in immunoassays require extremely sensitive methods of detection because of the very low molar concentration of the labelled species.

(*iii*) Anti-oxytocin is the antigen present in antiserum raised in order to determine oxytocin by FIA. ⟶

SAQ 5.3f (cont.)

(*iv*) Radioactive labels are are more likely to affect the immune reaction than fluorescent labels.

(*v*) Small analyte molecules may require linking to a protein molecule in order to provoke an immune response in a laboratory animal.

(*vi*) In a fluorescence enhancement method the free labelled antigen has a higher fluorescence efficiency than the bound antigen.

5.4. ROOM TEMPERATURE PHOSPHORESCENCE

It was pointed out in Part 1 that, in order to observe phosphorescence from the species present in liquid samples, it has generally been necessary to freeze them in liquid nitrogen in special solvents. This is very inconvenient and has in the past prevented phosphorescence catching on as a routine analytical technique. The introduction of instruments with pulsed sources for general fluorescence work has made it possible to provide limited time resolution facilities. This has led to a revival of interest in short-term phosphorescence and delayed fluorescence and stimulated a search for alternative procedures to allow phosphorescence to be observed at room temperature with samples of analytical interest. Two approaches look particularly promising, the use of solid samples and the protection of species in the liquid phase by the use of micelles, and room temperature phosphorescence (RTP) has now achieved the status of having widely recognised initials!

Π How do you account for the success with solid samples in this technique?

Samples in the solid state have long lifetimes even at room temperatures because the excited state cannot be deactivated by collision.

Π Why is scattering, which makes the observation of fluorescence difficult with solid samples, of no consequence in phosphorescence measurement?

Scattering ceases instantaneously when the source is switched off and so does not interfere with phosphorescence which is observed during the gaps between pulses.

Let's have a look at the possibilities of RTP with both solid and liquid samples.

5.4.1. RTP with Solid Samples

Π What experimental arrangement is normally required for the observation of photoluminescence from solid samples?

Frontal illumination as illustrated in Section 2.7.

Π Why is it not possible to use the 1 cm liquid cell as a container for powdered samples?

Incident radiation is strongly scattered by the crystal faces of the particles in a powdered solid which effectively prevents its penetration beyond the surface layers. 90° geometry could be used with large blocks of homogeneous materials such as clear plastics or glasses (luminescence standards are often supplied in this form).

The direct analysis of powdered samples in the solid state has much to commend it since it avoids what is often a time-consuming and labour intensive operation of bringing the sample into solution. Unfortunately problems arise with spectroscopic techniques in the optical region of the spectrum (uv, visible and ir) because of the fact that the dimensions of particles are comparable with the wavelength of the radiation and the results obtained are very dependent upon particle size, particularly when different components in a sample may exist in different ranges of particle size. These problems can be minimised by careful grinding of the sample and blending with inert material to achieve a homogeneous sample – but then we are back with a lengthy sample preparation and a potential advantage of solid sampling is lost. In general, therefore, it is better to carry out accurate quantitative analyses of solids using techniques such as X-ray fluorescence or nmr spectroscopy where the wavelength of the radiation is very different from the particle dimensions.

Π One possible way of making phosphorescence measurements on samples in solution is to evaporate them down to dryness. This can conveniently be done by placing a drop on a piece of filter paper which spreads out the sample and avoids high local concentrations of the solute material so that the analyte is present as very tiny particles trapped in the fibres of the paper. There are two potentially unsatisfactory features of this technique. What are they?

(*a*) There may be a high fluorescence background from the paper;

(*b*) Luminescence is recorded only from the surface layers of the paper.

Π Why may the first of these features not be troublesome with phosphorescence measurements?

The background fluorescence may decay very rapidly and not interfere with phosphorescence measured with some delay after the end of the pulse. It is also likely to be at much shorter wavelength than the phosphorescence.

Π What problem will arise as a result of the second feature?

Quantitative results may be erratic if concentration in surface layer is not representative of the entire sample due to uneven distribution.

Π Phosphorescence analysis becomes a particularly attractive proposition when the sample is already present on a filter paper or similar supporting medium. Can you think of an example where this might apply?

There are two common examples in routine analysis:

(*a*) Identification of spots separated by paper chromatography or tlc. We have already discussed this in Section 5.2.

(*b*) Analyses of mixtures trapped on filters during air pollution studies. The standard technique is to draw a large volume of the atmosphere to be tested through a glass fibre filter paper over a long period, typically an 8-hour working day. The filter is then removed and cut into pieces for analysis by different techniques. Metals are usually analysed directly on the support by XRF but most other analyses require the sample to be quantitatively extracted from the filter. This can raise practical problems and is generally a time-consuming process. RTP offers a possible method of in situ analysis for organic pollutants as well as elements not easily accessible by XRF such as beryllium and boron for which sample pre-treatment is minimal. Derivatisation can be carried out if necessary on the filter and, as before, the problems of background emission which might have plagued fluorescence are more easily avoided in phosphorescence measurements.

5.4.2. RTP with Liquid Samples

The problem of preventing the quenching of phosphorescence of compounds in the liquid phase can be tackled in ways other than by freezing them or evaporating them to dryness. One possible method is to increase the viscosity of the sample by the addition of a viscous compound such as glycerol (1,2,3-trihydroxypropane). This decreases the collision rate of the triplet state molecules in the same way as lowering the temperature (which also increases the viscosity of liquid samples) but the addition of glycerol is much more effective and convenient. A further advantage of glycerol is that it is completely miscible with water and can actually replace water or ethanol as a solvent for polar compounds. The phosphorescence of eosin can be observed at room temperature with glycerol as the solvent though the intensity is very much less than that of the prompt fluorescence. Eosin is a compound which also exhibits delayed fluorescence and this too can be observed at room temperature with solutions in glycerol.

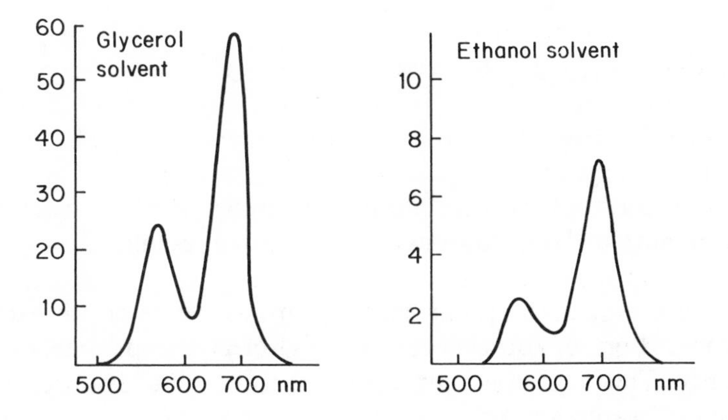

Fig. 5.4a. *Emission spectra of eosin in ethanol and in glycerol*

Π The phosphorescence and delayed fluorescence emission spectra of eosin in ethanol and glycerol at 20 °C are shown in Fig. 5.4a. Which band is which?

The phosphorescence band is at 690 nm and the delayed fluorescence band is at 580 nm.

Π On the same intensity scale the prompt fluorescence under the same conditions has an intensity of 10 000 units. What would its wavelength be?

580 nm, the same as the delayed fluorescence.

Very recently phosphorescence has been observed from molecules protected from de-excitation by collision using 'micelles'. This ingenious method makes use of the fact that when large organic molecules having a polar group at one end are placed in water they form spherical aggregates in which the long organic part of the molecule is directed towards the centre with the polar groups distributed over the surface. The most familiar example of this type of behaviour occurs with soaps which are sodium salts of long-chain fatty acids containing 12 to 18 carbon atoms. We have to introduce some new terms to describe this behaviour which you may not previously have met so let's be quite clear what they mean. The parent acid of a typical soap, stearic acid, is shown as

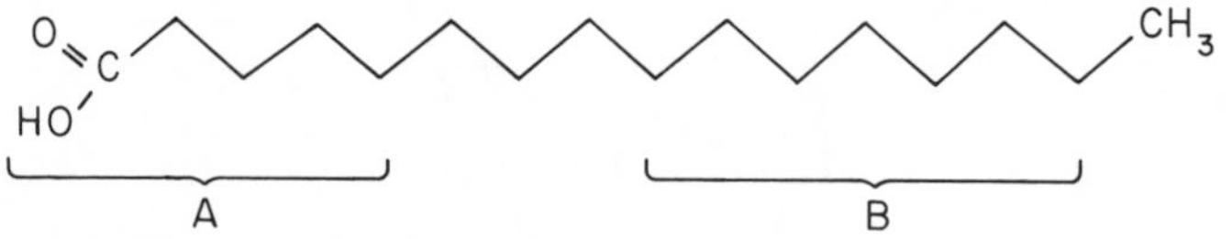

(Only the terminal carbon atoms are shown in this structure; you must imagine a CH_2 group at each kink in the chain.)

Region A contains the polar carboxyl group which confers upon the lower members of the carboxylic acid series the property of water solubility. This region is therefore said to be 'hydrophilic' from the Greek for 'water-loving'.

Region B is a non-polar hydrocarbon chain which is typically 'organic' and not attracted to water. This region is therefore said to be 'hydrophobic' (from the Greek for 'water-fearing') which is the direct opposite of hydrophilic. The word 'lipophilic' (meaning 'fat-loving') is often used in place of 'hydrophobic' – particularly by biologists. As the chain gets longer the hydrophobic (lipophilic) property dominates and the solubility decreases - stearic acid itself is virtually insoluble in water. However, the long-chain fatty acids do dissolve in alkali and in the aqueous medium the anions orientate themselves according to the 'like-with-like' principle. The lipophilic (hydrophobic) carbon chains seek out other organic material which, in this context, is limited to the carbon chains of other molecules. The carbon chains therefore get together and virtually dissolve in each other to produce what is in effect a small droplet of organic phase with polar groups embedded in the surface suspended in the aqueous phase and surrounded with an 'atmosphere' of Na ions. This is shown below. The droplet has a diameter approximating to twice the length of the carbon chain or about 4 nm (considerably less than the wavelength of uv radiation).

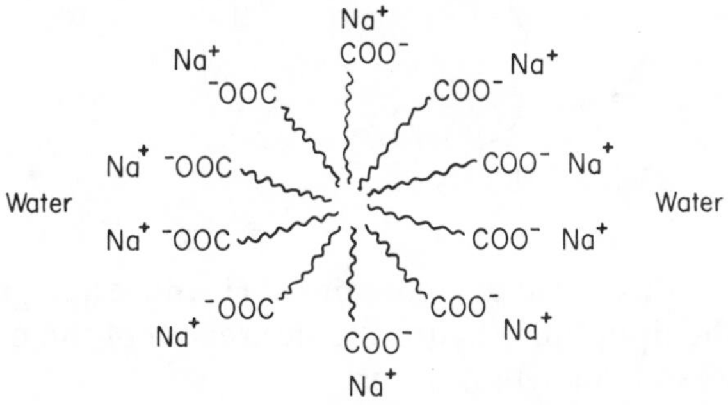

It is the presence of these micelles which gives to soap solutions the property of dissolving oils and grease. These are absorbed into the lipophilic region at the centre of the micelle. In a similar way, phosphorescent molecules can be absorbed into the centre of the micelle where they are protected from collision with water molecules, and, more particularly, oxygen molecules which are present in the aqueous phase.

Π A further development of this technique is to use the thallium(I) salt of the fatty acid rather than the sodium salt, for example thallium dodecyl sulphate. What benefit does this bring to phosphorescence methods?

Thallium is a 'heavy atom' (M_r = 204) which, as we have seen, will have the effect of enhancing phosphorescence emission.

Other large organic molecules have similar properties to micelles which enable them to provide a safe haven for triplet state molecules in aqueous media. The cyclodextrins, which are carbohydrates related to starch, have been found to be particularly successful for phosphorescence work. These compounds consist of a ring of D-glucose units which form a structure, variously described as a 'bracelet', a 'bucket' or a 'dough-nut' with hydroxyl groups above and below as shown here.

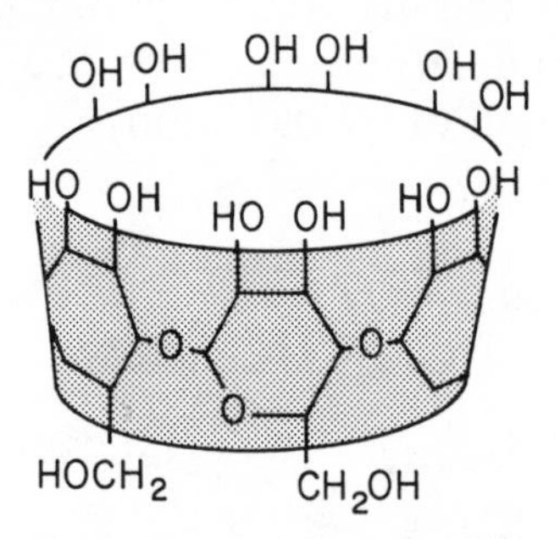

The inside of this structure contains CH and ester groups which constitutes the lipophilic (hydrophobic) region of the molecule into which the organic analyte is absorbed.

We shall shortly look at an example of the use of cyclodextrins in more detail as our case study in RTP.

SAQ 5.4a

Identify the phrase(s) which correctly complete the following sentences:

A Recent developments in RTP have taken place because

Phrases

(*i*) the previous approach used in phosphorescence of freezing samples in liquid nitrogen is inconvenient in analysis.

(*ii*) it avoids problems due to scattering from turbid samples.

(*iii*) analytes can be examined in situ on filters and paper chromatograms.

(*iv*) the use of pulsed sources and electronic gating has made phosphorescence measurement easier.

(*v*) it provides a method of analysis for samples which do not fluoresce.

B Solid samples can be successfully examined by RTP because

Phrases

(*i*) the technique is not affected by particle size.

(*ii*) scattering is excluded by observing the emission when the sample is not being irradiated. ⟶

SAQ 5.4a (cont.)

(*iii*) quenching by collision is prevented due to the lack of mobility in the solid phase.

(*iv*) frontal illumination can be used.

(*v*) no sample preparation is required.

SAQ 5.4b

Fill in the blanks in the following paragraph using words or phrases selected from the list below:

Phosphorescence may be observed from fluid solutions at room temperature by taking steps to reduce the .... between analyte molecules and potential .... such as dissolved oxygen. This can be done by .... the viscosity of the solution by .... the temperature or adding a viscous compound such

⟶

SAQ 5.4b (cont.)

as .... Alternatively, the excited analyte molecule can be protected by absorption into the .... region of a .... formed by the .... of long-chain fatty acids, particularly in the presence of heavy cations such as .... which facilitate .... to the triplet state. Certain types of organic molecules such as .... have a similar structure and behave in the same way.

hydrophilic
increasing
inter-system crossing
rate of collision
cyclodextrin
glycerol
glucose
potassium
thallium
lipophilic
lowering
quenching agents
ethanol
antigen
anions
derivatising agent
micelle

5.5. ANALYSIS OF MIXTURES OF THE PRINCIPAL OPIUM ALKALOIDS

This method, taken from a relatively old paper by Chalmers and Wadds [*Analyst 1970*, **95**, 234], illustrates very clearly the use of natural fluorescence modified by control of pH and solvent extraction to make the method more specific. The analytes concerned are still of great interest as 'drugs of abuse' and you can easily imagine circumstances where this method might be used.

We are interested in the four alkaloids, morphine, codeine, papaverine, and narcotine. Their structures are

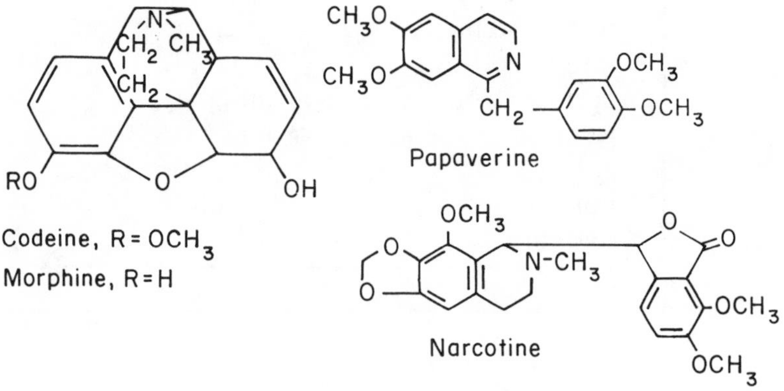

Codeine, R = OCH_3
Morphine, R = H

Papaverine

Narcotine

Π What features do they have in common?

All contain at least one fused ring system and, in every case, one of the rings contains a nitrogen atom which is responsible for the basic properties of the molecules.

We shall first consider the chemical and spectroscopic principles involved in the method and then work our way step by step through the entire procedure to a final calculation based on some typical experimental data. This will, inevitably, involve you in a mass of detail so you should make a list of the essential points as we go along – then you won't get lost or bogged down! In fact, this is exactly what you would do if you were carrying out the analysis for real in the laboratory – assuming, of course, that you have been properly trained in the use of the 'laboratory notebook'!

5.5.1. Basis of the Method

Codeine and morphine can be determined independently because, while both fluoresce strongly at the same wavelength in sulphuric acid (0.05 mol dm^{-3}), morphine has a negligible fluorescence in sodium hydroxide (0.1 mol dm^{-3}). The calibration curves for both compounds in both solvents are shown in Fig. 5.5a. The emission is measured at 345 nm with excitation at 285 nm. Intensities are

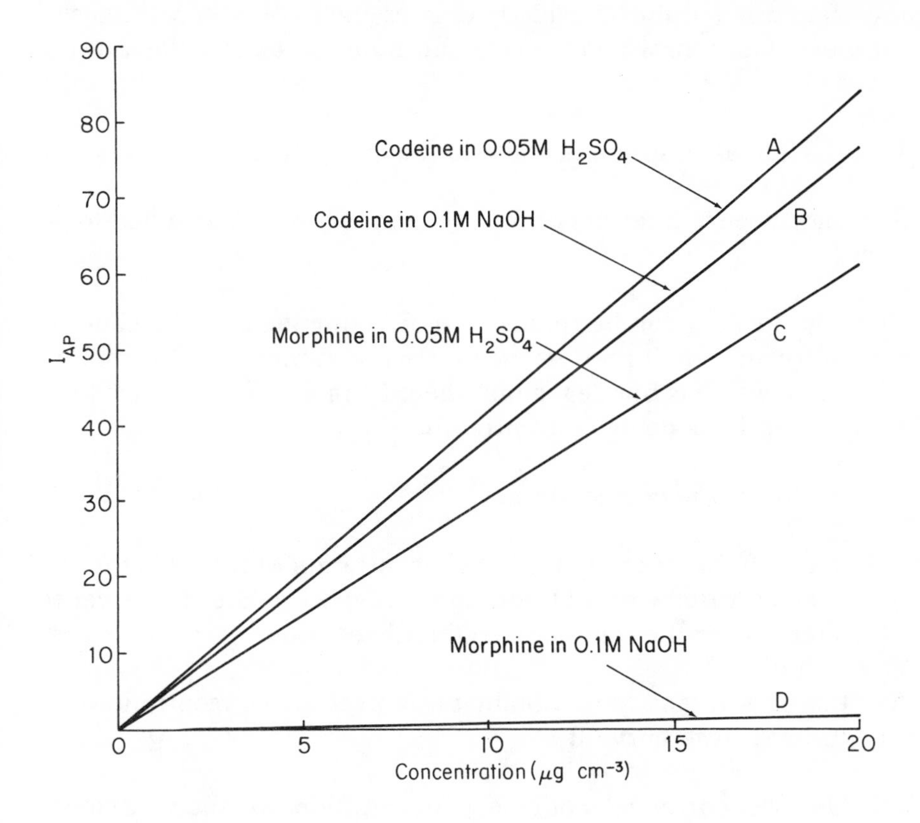

Fig. 5.5a. *Calibration graphs for codeine and morphine:*

(*A*) Codeine in H_2SO_4 (0.05 mol dm^{-3})
(*B*) Codeine in NaOH (0.1 mol dm^{-3})
(*C*) Morphine in H_2SO_4 (0.05 mol dm^{-3})
(*D*) Morphine in NaOH (0.1 mol dm^{-3})

expressed as a percentage of the intensity for a standard solution of 2-aminopyridine (1.0 mg dm^{-3}) in sulphuric acid (0.05 mol dm^{-3}) measured at the same wavelengths.

The codeine is first determined by the fluorescence of the sample in NaOH solution and using curve B. The fluorescence intensity of this concentration of codeine in sulphuric acid is then calculated from curve A and subtracted from the measured fluorescence intensity of the sample in sulphuric acid. This gives the fluorescence intensity for morphine from which the concentration can be calculated using curve C.

Π What assumption is made in this procedure?

The fluorescence intensity of the two compounds are assumed to be additive.

The initial values can be refined by correcting the codeine concentration for the small fluorescence intensity of morphine in sodium hydroxide which is not negligible when the morphine concentration is high and the codeine concentration is low.

Π How is the correction made?

(*a*) Read off the intensity for the concentration of morphine in the mixture from curve D and subtract it from the fluorescence intensity for the sample in sodium hydroxide;

(*b*) Use this intensity to obtain the corrected concentration for codeine from curve B;

(*c*) Use the corrected codeine concentration to obtain a more accurate value for the concentration of morphine.

Papaverine and narcotine do not interfere seriously with the determination of codeine and morphine because they fluoresce at longer wavelengths. However they both absorb significantly at 285 nm (the excitation wavelength for codeine and morphine) and the accuracy of the determination can be improved by extracting them from the sulphuric acid solution with ethanol-free chloroform.

Π Why does this procedure improve the accuracy?

Absorption by other components in the solution constitutes an inner filter effect which reduces the intensity of fluorescence of the analytes. This would not matter if the concentrations of papaverine and narcotine were fixed because the effect would be constant and could be allowed for by matrix-matching the standards. However, in practice the concentrations will vary from sample to sample and so the effect is not predictable. In any case, the solvent extraction is probably simpler than matrix-matching the standards and leads to an enhancement in the sensitivity which brings with it a gain in signal/noise ratio.

The extraction of papaverine and narcotine into chloroform can be made quantitative by adjusting the pH of the aqueous phase to 9 with a borate buffer solution. This is the first step in the determination of these two alkaloids which are also separated from codeine and morphine. In chloroform, papaverine fluoresces strongly at 348 nm when excited at 320 nm. Under the same conditions, narcotine fluorescence is very weak and occurs at much longer wavelength, 400 nm, where it does not interfere with the determination of papaverine.

To measure the narcotine a rather cunning device is used. The addition of a small amount of trichloracetic acid (trichlorethanoic acid, CCl_3COOH) is found to enhance the fluorescence of narcotine by a factor of more than 10 and shift it to shorter wavelength (λ_{ex} = 315nm, λ_{em} = 375nm). At the same time the fluorescence of papaverine is considerably suppressed and shifted to longer wavelength (λ_{ex} = 417nm, λ_{em} = 455nm) so that it does not interfere with the determination of narcotine. So we are able to measure the concentration of both alkaloids independently in the chloroform extract and complete the analysis of the four-component mixture. Excitation and emission spectra of papaverine and narcotine in chloroform with and without the addition of trichloracetic acid are shown in Fig. 5.5b.

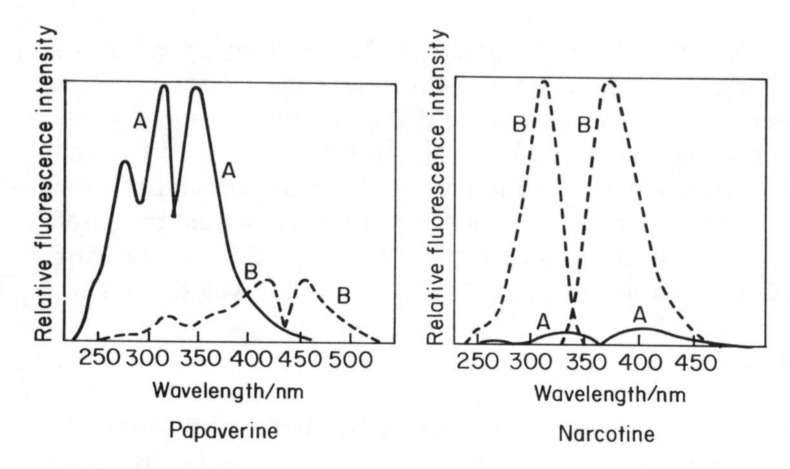

Fig. 5.5b. *Spectra of papaverine and nacroctine*

5.5.2. Laboratory Procedure

Let's now imagine that we are in the laboratory and about to analyse a sample of papaveretum, a commercial mixture containing the four alkaloids in question as their hydrochlorides and used as the basis of some analgesic preparations. First of all we shall have to assemble the apparatus and chemicals needed.

(*a*) *Instrumentation*

We shall need a dual monochromator spectrofluorimeter in order to select the correct wavelength for each compound, (a filter instrument could be used if we had the right filters but this would have required advanced notice if we had not previously used this method). The authors of the original paper used a Farrand instrument with a 150 watt xenon arc source, an RCA IP28 photomultiplier detector and both a meter and a 15 inch chart recorder as read-out. This was in 1970 of course, but any of the currently available instruments would do just as well. The authors used a band-width of 10 nm which is a standard setting on modern instruments. We shall also need a pair of good quality 1 cm silica cells (the authors used quartz).

Π Why not glass?

The excitation wavelength for codeine and morphine is 285 nm which is below the usable transmission limit for glass (320 nm)

(*b*) *Reagents*

The first chemicals we need are pure samples of the four alkaloids and 2-aminopyridine. These can be readily obtained from commercial sources but they will probably need to be purified by recrystallisation until they give a constant melting point and their uv spectra conform to published data.

We shall also need pure trichloracetic acid which may also need to be recrystallised from chloroform and dried over phosphorus pentoxide. Once we have our pure chemicals we can set about preparing the reagents we need. Let's make a list of them.

(*i*) Sulphuric acid (0.05 mol dm^{-3})

This can be bought ready made up and standardised but it would be a waste of money to use this since we do not need the concentration to be exactly 0.05 mol dm^{-3}. We are not going to titrate the acid or use it in a quantitative process so as long as we use somewhere near 0.05 mol dm^{-3} that will be good enough. It is very important in any analysis to distinguish carefully between measurements which have to be made to the highest possible precision and those which do not. This is one of the latter. We shall of course need a good quality grade of the acid with low fluorescent background which may entail testing one or two of our stock bottles obtained from various sources. Once we are satisfied with the stock we take 5.6 cm^3 and dilute it to 2 litres with good 'fluorescence quality' distilled water.

(*ii*) Sodium hydroxide (0.1 mol dm^{-3})

The same remarks apply as to sulphuric acid. We shall need to weigh out 8.0 g of analytical reagent grade sodium hydroxide, dissolve it in distilled water and make up to 2 litres.

(*iii*) 2-aminopyridine standard solution

This is not very stable so if we are likely to be making measurements over a period of days or weeks (for instance if we were going to use the method routinely for quality control of a manufacturing process) we shall need to make up a stock solution by dissolving 0.1000 g of the pure solid in sulphuric acid (0.05 mol dm^{-3}) and diluting it to 1 litre. This is then stored in a refrigerator and working solutions containing 1.00 mg dm^{-3} are made up as required by diluting 10 cm^3 of the stock solution to 1 litre.

(*iv*) Ethanol-free chloroform

AR-grade chloroform usually has an acceptably low fluorescence and if we are lucky we shall be able to use it 'as received'. If the fluorescence amounts to more than 0.3% of the 2-aminopyridine standard however it will be necessary to redistil it using all-glass apparatus.

Chloroform is always sold with about 2% of ethanol added as a stabilizer and, if not removed, this would impair the efficiency of the solvent extraction process. To remove the ethanol, chloroform is shaken twice with an equal volume of distilled water on each occasion. It is then filtered through a plug of anhydrous sodium sulphate and stored in the dark.

(*v*) pH 9 buffer solution

Commercial buffer tablets are available but it is safer to make this solution up from good quality starting materials. It is very simply prepared by adding hydrochloric acid (0.1 mol dm^{-3}, 46.0 cm^3) to sodium borate solution (0.025 mol dm^{-3}, 500 cm^3).

(*vi*) Trichloracetic acid

We shall need a 1% w/v solution in our low fluorescence AR grade chloroform (ie 1 g in 100 cm^3).

(*c*) *Calibration Standards*

Π We are now in the position to proceed with the with the preparation of the standard solutions of the four alkaloids. We need to make up a stock solution of each of the alkaloids from which to prepare a set of working standards to cover the range 0.1 to 20.0 mg dm^{-3}. What concentration shall we make the stock solution, how much shall we make up and what shall we use as the solvent?

A suitable concentration would be 1 g dm^{-3} and we should require 100 cm^3 of each stock solution. Codeine and morphine would be made up in distilled water and papaverine and narcotine in ethanol-free chloroform. You may have come up with different answers so let's have a look at the principles on which these values are based. You will then see if your proposals are equally valid.

Perhaps the most crucial factor is the weight of each solid to be taken. If this is too low the precision of the method will be impaired, whilst if we use larger quantities to improve the precision we shall have to hold relatively large stocks of what are quite expensive chemicals. (There is also the problem of security since these are dangerous drugs which must not be allowed to fall into the wrong hands.) To make 100 cm^3 of a 1 g dm^{-3} solution requires 0.1 g of solid which we shall be able to weigh out to the nearest 0.0005 g on our analytical balance, a precision of $\pm 0.5\%$. The solvent is determined by the fact that codeine and morphine are measured in the aqueous form and papaverine and narcotine in chloroform. Of course, the latter could be dissolved in water and extracted into chloroform as in the procedure for the mixture but this would increase the time taken to do the calibration (though it might also improve the accuracy). Codeine and morphine could have been dissolved in sulphuric acid (0.05 mol dm^{-3}) or sodium hydroxide (0.1 mol dm^{-3}) since calibrations are required in both solvents. Ideally it would be best to have a stock solution in each to avoid altering the concentration during the dilution stage but this is not likely to cause sufficient error to merit the additional trouble.

Π What working standards shall we make up to cover the required calibration range (0.1 to 20 mg dm^{-3})?

A suitable set would be 20, 16, 12, 8, 4, 1, 0.4 mg dm^{-3}. Again we have to balance the needs of precision against the time and expense (time is money in a commercial laboratory) of making up and running a large number of standards. This set provides adequate precision in the middle and top of the range where it is likely to be of greatest importance. The calibration range is more than two decades (factor of >100) and if it is necessary to achieve equal precision at the bottom end it will be necessary to include additional standards below 1 mg dm^{-3} and re-plot the range from 0–1 mg dm^{-3} on a larger scale.

Π How shall we make up these standard solutions from the stock solution?

Π How many sets of standards shall we be able to make from the whole batch?

First dilute the stock solution by taking 10 cm^3 and making up to 100 cm^3 in the appropriate solvent. Then take 10, 8, 6, 4, 2, 0.5 and 0.2 cm^3 aliquot portions of the diluted stock and make up to 50 cm^3 for each standard. Each calibration therefore requires 30.7 cm^3 so the diluted stock is enough for 3 calibrations. Hence the whole stock solution will do 30 calibrations. In the case of codeine and morphine where we have to calibrate in both sulphuric acid and sodium hydroxide there is enough for only 15 complete sets of calibration standards.

Π Once again you may have made a different selection but if so it is likely to be a less good solution to the precision versus cost compromise. What two disadvantages would arise from the procedure of taking 2.0, 1.6, 1.2, 0.8, 0.4, 0.1, and 0.04 cm^3 of the original stock and making up to 100 cm^3?

If we use a micro-burette or a graduated pipette, the precision of measurement of the aliquot portions is less, particularly at the lower end of the range below 0.5 cm^3. It would also consume much more solvent (700 cm^3 instead of 450 cm^3). Thus we lose out on both counts! The amount of solvent is significant since, apart from the cost of AR-grade chloroform, we have made up only 2 litres of the sulphuric acid (0.05 mol dm^{-3}) and sodium hydroxide (0.1 mol

dm^{-3}). This is enough for only one calibration since each is used with two of the alkaloids. We shall also need some of the solvents for the samples and we shall have to provide a blank for each of the calibrations and carry a blank through the sample procedure.

Modern micro-pipettes with a syringe action are far more precise than the traditional glass pipettes and if we use one of these we can maintain a precision of $\pm 1\%$ down to 0.01 cm^3 so the first objection is no longer valid – though the volume of solvent is still a drawback. However, if micro-pipettes are available, we could reasonably take half these quantities (1.0 to 0.02 cm^3) and make up to 50 cm^3 and consume the same amount of solvent as in the second stage of the double dilution procedure.

(*d*) *Preparation of the Sample Solutions*

You are by now, no doubt, anxious to start making fluorescence measurements and plotting the calibration graphs but don't be in too much of a hurry! We need to prepare our samples first so that they can be measured at the same time as the standards. This is both convenient and more accurate since it avoids the possibility of any 'drift' in instrument response between running the standards and the samples. In fact, this should not be too much of a problem with the present method because it uses a fluorescence intensity standard (the 2-aminopyridine solution) to compensate for any instrumental variations. However, it is always good practice to run samples and standards as close together as possible.

We should now switch the instrument on so that it will have thoroughly warmed up and stabilised when we come to use it. If there is likely to be a lengthy delay while we prepare any of the samples it would be as well to defer switching on the source until we are about 30 minutes away from making measurements. The life of high intensity xenon arcs is limited and it is as well not to waste it unnecessarily. In some instruments the source can be switched on and off independently.

Our sample will be in the solid state and the first step is to get it into solution. A sample weight of 40–50 mg is suggested which is dissolved in sulphuric acid (0.05 mol dm^{-3}) and made up to 100

cm^3. This weight is rather low for an ordinary analytical balance and will reduce the precision by a factor of about 2 compared with the weight of standard that we used but this is likely to ensure that the concentration will ultimately lie within the calibration range for all four alkaloids. In some situations the use of such a small sample can be an advantage – in forensic work for example where the sample is limited to small spillage or residues. If precision is important it is possible to use a more accurate microbalance that will weigh to 0.01 mg.

The determination of codeine and morphine are done in the aqueous phase so let's start with that. We pipette 10 cm^3 of the solution into a separating funnel and shake it with 15 cm^3 of ethanol-free chloroform for one minute. When the phases have separated we can run off the chloroform (lower layer) and discard it. That gets rid of the papaverine and narcotine. We now have to run the aqueous phase into a 50 cm^3 flask and make the transfer quantitative by washing through with distilled water, finally making up to the mark.

Π How does the concentration of this solution compare with the original sample?

It is 5 times less. 10 cm^3 was taken, extracted with chloroform and then diluted to 50 cm^3.

We now pipette two 10 cm^3 aliquots of the extracted aqueous phase into 100 cm^3 flasks making one up to the mark with sulphuric acid (0.05 mol dm^{-3}) and the other with sodium hydroxide (0.1 mol dm^{-3}). We must take care to label these solutions clearly. Let's call the first A, and the second B (= acid and base!).

Π How do the concentrations of the alkaloids in A and B compare with those in the original sample solution?

They are 50 times less. The dilution factor of 5 in the extraction step has been increased by a further factor of 10 by diluting a 10 cm^3 aliquot to 100 cm^3 giving an overall factor of $5 \times 10 = 50$.

Now for the papaverine and narcotine which we have to extract into chloroform at pH 9. Chlorinated solvents are very toxic and so we must carry out this procedure in an efficient fume cupboard. We pipette 5 cm^3 of the sample solution into a separating funnel and add 10 cm^3 of our pH 9 buffer solution. We add 20 cm^3 of chloroform, shake for about a minute and run off the chloroform (lower) layer through a plug of anhydrous sodium sulphate moistened with chloroform into a dry 50 cm^3 standard flask. We must repeat this operation with a fresh 15 cm^3 of chloroform and again with a further 10 cm^3. The plug is then washed through with more chloroform into the flask and the volume made up to the mark. We shall label this solution P. We shall also need a blank made up in the same way but starting with 5 cm^3 of distilled water. We shall label this BP.

Π Why don't we extract the alkaloids with 45 cm^3 of chloroform in one go?

The multiple extraction procedure is much more efficient than using the same volume of solvent in a single extraction. It is necessary here to ensure quantitative extraction. A single extraction was enough for the codeine/morphine procedure because we were concerned only to remove the bulk of the papaverine and narcotine to minimise their interference and quantitative extraction was not necessary.

Finally we prepare the solution for determining narcotine by pipetting 25 cm^3 of solution P in a 50 cm^3 standard flask, adding 2 cm^3 of the trichloroacetic acid solution, and making up to the mark. We label this solution N and a blank solution, obtained by taking 25 cm^3 of solution BP through the same procedure, BN.

At last we have all the solutions and we can start measuring their fluorescence. Let's just make a list of them to remind ourselves of what we have done – you may already have made a similar list in your notes so you can now check it.

(*i*) Calibration standards of concentration 20, 16, 12, 8, 4, 2, and 0.5 mg dm^{-3} for:

(*a*) Codeine in sulphuric acid (0.05 mol dm^{-3})
(*b*) Codeine in sodium hydroxide (0.1 mol dm^{-3})
(*c*) Morphine in sulphuric acid (0.05 mol dm^{-3})
(*d*) Morphine in sodium hydroxide (0.1 mol dm^{-3})

(*e*) Papaverine in chloroform
(*f*) Narcotine in chloroform containing trichloracetic acid (0.04% w/v)

(*ii*) Blanks for calibration:

(*a*) Sulphuric acid (0.05 mol dm^{-3})
(*b*) Sodium hydroxide (0.1 mol dm^{-3})
(*c*) Chloroform
(*d*) Chloroform containing trichloracetic acid (0.04% w/v)

(*iii*) Samples:

(*a*) A, Aqueous sample in sulphuric acid (0.05 mol dm^{-3}) extracted with chloroform

(*b*) B, Aqueous sample in sodium hydroxide (0.1 mol dm^{-3}) extracted with chloroform

(*c*) P, chloroform extract from aqueous sample buffered to pH 9

(*d*) N, chloroform extract from aqueous sample buffered to pH 9 containing trichloracetic acid (0.04% w/v)

(*e*) BP, Blank chloroform extract

(*f*) BN, Blank chloroform extract containing trichloracetic acid (0.04% w/v)

5.5.3. Measurement of Fluorescence

Now to the spectrofluorimeter which has been switched on for the last half hour or so which is long enough for the instrument to settle down. The laboratory technician has provided us with a pair of clean 1 cm silica cells, all four sides polished.

Π Before we measure the fluorescence intensity of the aqueous solutions it is necessary to bubble oxygen-free nitrogen through them for 10 minutes. What is the reason for this?

To remove the dissolved oxygen. Dissolved oxygen causes the reading for morphine in particular to increase as you measure it because of photodecomposition. This step is a considerable nuisance since it means waiting for 10 minutes before taking each reading. It would therefore take about 1.5 hours to complete each calibration. This can be reduced by using a batch of, say, six bubblers to purge six solutions simultaneously.

Π There is a problem with using bubblers in quantitative analysis which is more serious when a batch is used. What is it?

Some evaporation of the solvent will occur, particularly if the temperature in the laboratory is high. This will increase the concentration of the analyte. If the same bubbler is used every time and the nitrogen flow rate is fixed this effect will not be serious since the reduction of volume will be constant. With a batch system however the flow rate is bound to be different for different bubblers because it is very difficult to make the dimensions of all the bubblers identical. This will result in a 'scatter' of the experimental values. In accurate work it would be necessary to check this factor carefully.

Matters would be worse with chloroform which is more volatile than water. Although the paper states that the calibration graphs for the chloroform solutions are obtained in a similar way (which implies purging with nitrogen) there is no mention of a problem with papaverine or narcotine due to dissolved oxygen which, in any case, is at a much lower concentration in the organic solvent.

Π Let's start with the codeine in sulphuric acid. First we must set the correct wavelengths and band-width. What values do we choose? (Refer to Section 5.5.1 and your notes.)

Excitation wavelength 285 nm Emission wavelength 345 nm

Band-width 10 nm

Π We have two cells. Which solutions shall we put in them?

Let's put the 2-aminopyridine standard in one and keep it there for the entire set of measurements. We will use the other for the standards and samples. It will be convenient to start with the 20 mg dm^{-3} standard.

Π Why should we choose these two solutions and why keep one cell specifically for the fluorescence standard?

We need the fluorescence standard to provide a fixed basis for the intensity scale so we shall need to use it to set the sensitivity of the instrument and to check it frequently throughout the measurements. If we run the top standard we shall check whether we can use the same setting for the calibration standards as the fluorescence standard. Ideally we should like the 20 mg dm^{-3} standard to give a reading close to that of the fluorescence standard. It is best to stick to the same cell for the 2-aminopyridine to avoid the risk of contaminating the cell in which low levels of fluorescence from blanks or low concentration samples might be measured.

The cell must be carefully washed out to remove all traces of the last solution before refilling with the next. This is most conveniently done by filling and emptying the cell with each new solution 3 or 4 times before measuring the fluorescence. The possibility of contamination from previous samples is also reduced if we take standards either in descending or ascending order to minimise concentration changes. It will also improve the accuracy if we take a second reading on each solution after emptying each cell and refilling it.

Π Right! Let's load the cells. In each case we fill the cell and empty it 3 times before we start readings, exactly as we shall do between readings. Before we place the cells in the instrument we shall close the excitation shutter. Why?

To minimise exposure of the solution in the sample position to uv radiation to avoid photochemical decomposition. Most instruments provide a 'turret' for holding up to 4 cells. Let's put the fluorescence standard in the sample position and the 20 mg dm^{-3} codeine standard in the next position.

Let's assume that we have a modern spectrofluorimeter with a

3-figure digital readout 0 to 999. With both the excitation shutter and the cell compartment lid closed it should be reading zero. If not we must adjust the zero control until we get 000. (With a modern instrument there will probably be an 'auto-zero' key which will do this for us.) Now, with the 2-aminopyridine standard in the sample position we open the excitation shutter and get a reading on the meter. We adjust this to, say, 900 using the sensitivity (gain) control and then close the excitation shutter. To check the reading, we shall empty the cell, refill it with fresh 2-aminopyridine standard, open the excitation shutter and adjust the reading to 900 if there is any change (which would indicate contamination of the previous sample).

Now we measure the fluorescence of the 20 mg dm^{-3} codeine standard. Let's suppose we get a reading of about 800. This is likely to be the highest value we shall get, at least with the aqueous solutions, so it is convenient that the reading is at the top end of the scale and close to that for the fluorescence standard. It should ensure that we maintain 3-figure precision over the whole range of standards. We renew the standard solution and take a duplicate reading.

We can now run through the other six standards and the blank, taking duplicate readings in each case, following the 'triple wash' routine throughout and finishing with solution A. We should also be wise to check the fluorescence standard from time to time, recording the updated reading at the appropriate point in the sequence.

Π To plot the calibration graph, we take the mean of each pair of duplicate readings and convert these to intensities relative to 100 for the fluorescence standard (taking the reading closest to the particular codeine standard in the sequence if there has been any change in the fluorescence standard). If the reading for the top calibration standard is 800 and the reading for the fluorescence standard is 900, what intensity value do we assign to the top standard?

88.9 ($= 100 \times 800/900$). Note that we quote 3 significant figures.

We convert the readings on the other standards, the blank and solution A in a similar way and subtract the blank from all the other values and plot the calibration graph using the intensity values for the standards. It is good practice to plot these values, roughly at least, straight away so that we can see if any point is seriously out of line and repeat the reading while the standards are still to hand and the instrument set up. If necessary, we may have to remake the standard. Our calibration graph should look like curve A in Fig. 5.5a.

Π The morphine standards in sulphuric acid are run next following the same procedure as for codeine. The same sulphuric acid (0.05 mol dm^{-3}) blank is used and, of course, we do not need to run solution A again. Do we need to change any of the instrument settings?

No. The same wavelengths are used for morphine and codeine and the band-pass is 10 nm throughout. According to Fig. 5.5a we should get significantly lower values for morphine (curve C) but the same sensitivity setting will give us adequate precision.

The corresponding curves for the sodium hydroxide solutions (B and D) are obtained by a similar series of readings on the appropriate sets of standards (B and D) using sodium hydroxide (0.1 mol dm^{-3}) as blank. The reading for Solution B is also obtained. Again, no change is required to the instrumental conditions. We could increase the precision of the values for the morphine standards since they will be very low but this is not worth the trouble since they are used only to correct the codeine concentration.

That takes care of codeine and morphine. We have the calibration graphs and intensity values for Solutions A and B so we can calculate the concentrations of these alkaloids in the mixture by the method described earlier. (We'll have a go at an actual example in an SAQ in a moment.)

Π Describe the procedure for determining the papaverine content of the mixture.

(*a*) Set the excitation and emission monochromators to the appropriate wavelengths (320 and 348 nm respectively).

(*b*) Place the fluorescence standard in the sample position, set the sensitivity to the same value used previously, open the excitation shutter and take the reading. (It will, of course, be different from the previous value because we have changed the wavelengths.)

(*c*) Run through the set of papaverine standards in duplicate using the triple wash procedure as before and read the intensities of Solution P and the blank BP in the same way. No intensity data are quoted in the paper so we shall assume that we can use the same sensitivity as for codeine and morphine.

(*d*) Plot the calibration graph for papaverine and read off the concentration of solution P remembering to subtract the blank from all readings before converting to the intensity value relative to 2-aminopyridine. Refer the concentration back to the mixture to obtain the papaverine content as % w/v in the solid.

A similar procedure is followed with the appropriate wavelength settings for the narcotine standards using solution BN as the blank. The concentration of narcotine in the solution is obtained from the reading for sample N and, from this, the concentration of narcotine in the mixture can be calculated.

It has taken us very many pages to describe this assay in detail which may give the impression that it is a very lengthy procedure in practice. Of course, it will take some time to set up the method – preparing standard solutions, optimising the conditions and preparing the calibration graphs and the like. However, once that is done, the authors claim that duplicate analyses for the four alkaloids can be carried out in 2.5 hours. They also claim that interference from other compounds present in opium is negligible.

SAQ 5.5a

0.0453 g of papaveretum (an opium concentrate) was weighed out, dissolved in sulphuric acid (0.05 mol dm^{-3}) and made up to 50 cm^3. A 10 cm^3 aliquot of this solution was trans-

⟶

SAQ 5.5a (cont.)

ferred to a separating funnel and shaken with dry, ethanol-free chloroform (15 cm^3) for one minute. When the phases had separated, the chloroform (lower) layer was discarded and the aqueous phase was washed quantitatively into a 50 cm^3 standard flask with distilled water and made up to the the mark. Separate 10 cm^3 aliquots of this solution were transferred to two 50 cm^3 standard flasks. One of these was made up to the mark with sulphuric acid (0.05 mol dm^{-3}) and labelled 'solution A'. The other was made up with sodium hydroxide (0.1 mol dm^{-3}) and labelled 'solution B'. After purging the solutions, the blanks and the 2-aminopyridine standard with oxygen-free nitrogen for 10 minutes, the following fluorescence readings were obtained with the excitation monochromator set to 285 nm and the emission monochromator to 345 nm:

Solution A	574
Solution B	112
Sulphuric acid (0.05 mol dm^{-3})	27
Sodium hydroxide (0.1 mol dm^{-3})	43
2-aminopyridine standard	924

(*i*) Calculate the percentage of codeine and morphine in the sample using the calibration curves in Fig. 5.5a.

(*ii*) Which is the least accurate value in this calculation?

(*iii*) Is three-figure accuracy justified?

(*iv*) What is the purpose of ratioing the readings on the samples to that for 2-aminopyridine in this method?

SAQ 5.5a

5.6. DETERMINATION OF SELENIUM IN NATURAL WATERS

Let's now take an inorganic analysis, involving one of the 'metalloid' elements. These elements, which include arsenic, antimony and boron as well as selenium and tellurium have always been troublesome in analysis – even by atomic absorption spectroscopy which, it must be admitted, is likely to be the preferred technique for metals unless the level is very low. The technique of hydride generation has in recent years improved the detection limits for these elements though it is much less convenient than 'straight' AAS and not without its problems. Plasma emission spectroscopy probably supplies the best answer but only at very considerable expense (£30 000 or more). Spectrofluorimetry, however, still has much to offer in the case of selenium and, though the method involving the formation of a derivative with 2,3-diaminonaphthalene (DAN) was first reported in 1962, several papers have appeared in the recent literature.

The method we shall follow was published in 1983 by Takayanagi and Wong [*Anal. Chim. Acta*, 1983, **148**, 263] for use in oceanographical surveys, and demonstrates the use of preconcentration to enhance the sensitivity of an analytical technique and the use of the method of standard additions. It also permits separate determination of the element in different valency states, Se(IV) and Se(VI), to be made.

Π The total selenium concentration in sea-water falls in the range 0.3–2.0 nmol dm^{-3}. What is this in μg dm^{-3}?

[A_r Se = 79.0]

0.024–0.16 μg dm^{-3}.

5.6.1. Basis of the Method

The selenium(IV) is first complexed with ammonium 1-pyrrolidine dithiocarbamate (APDC) at pH 4.2 and extracted into chloroform. This both provides a high degree of preconcentration and separates Se(IV) from Se(VI). The selenium is then back-extracted from the chloroform layer into nitric acid (8 mol dm^{-3}) and the solution is evaporated to near dryness after adding a small quantity of perchloric acid (70% w/v).

The DAN complex is then formed at pH 1, after first adding EDTA, and extracted in cyclohexane. Its fluorescence is excited at 380 nm and measured at 520 nm using band-widths of 4 nm and 10 nm respectively. A Perkin–Elmer MPF-44 spectrofluorimeter was used but their is no reason why this analysis should not be carried out with a simple filter fluorimeter fitted with interference filters having the appropriate specification.

Π What is the purpose of adding EDTA?

It is probably a masking agent to form complexes with other elements which might also form complexes with APDC and be extracted along with the selenium.

The procedure for total selenium involves precipitation as elemental selenium by reduction with hydrazine sulphate in the presence of tellurium which acts as a carrier. The tellurium is formed by a similar reduction of excess sodium tellurite and the precipitate is filtered off, dissolved in nitric acid containing a small amount of perchloric acid (70% w/v) and evaporated to near dryness. The DAN complex of selenium is formed and its fluorescence measured as before (tellurium does not interfere).

Π Explain the function of tellurium as a carrier.

The sodium tellurite is reduced to tellurium by the hydrazine sulphate and precipitates out carrying with it the selenium reduced from the selenite present. The function of a carrier is to ensure that there is a reasonable amount of precipitate to filter. The trace amount of selenium alone would not be visible and would easily be lost.

Π The selenium content is determined by the method of standard additions in which known amounts of selenium are added to aliquots of the sea-water sample which are then taken through the above procedures. Explain how the selenium content is derived from the fluorescence readings in this process.

The fluorescence reading is plotted against the concentration of selenium added to the aliquot. The graph is then extrapolated back until it cuts the concentration axis. The (negative) intercept on this axis represents the concentration of selenium in the aliquot with zero addition (ie in sea-water itself).

Much of the work reported in the paper is devoted to validating and optimising the procedure for sea-water samples. The separation steps in the preconcentration stage (extraction of the APDC complex into chloroform and coprecipitation of of selenium with tellurium) are particularly crucial to the overall accuracy of the method and were tested by carrying out the full procedures on standard solutions of known selenium content in deionised water. These tests showed the procedure to be quantitative for both selenium(IV) and total selenium though the precision was no better than $\pm 5\%$. However, in trace analysis high precision is not usually of great importance, the main aim being to establish reliably whether the level of a particular species is above or below its toxic limit. In the case of selenium, the permitted limit is 0.1 $\mu g\ dm^{-3}$ and it makes no difference whether the actual value is 0.236 or 0.237 $\mu g\ dm^{-3}$ when the analytical results are used in legal proceedings against a defendant accused of dumping toxic waste –he's going to be in trouble either way!

Besides this, in many environmental and biological applications of analytical chemistry the aim is to establish a trend or a departure from the norm in a situation where the natural variation in the level of the analyte is quite large. In these circumstances the accuracy of a particular measurement is of secondary importance provided we know what the limits are; $\pm 10\%$ is quite adequate in most cases.

5.6.2. Laboratory Procedure

Let's now put the method into context in the laboratory. The authors were concerned in establishing the selenium concentration at various depths in the Sargasso Sea. Unfortunately our samples are likely to be rather less exotic though the method will be much the same. Let's suppose that we have been asked to analyse the water in a local river for selenium as part of a general investigation of pollution in the area. The sample was collected by well-trained staff from our own laboratory who appreciated the need to acquire a representative sample of the river which would be large enough to allow for a high degree of preconcentration and several replicate determinations. They have come back with 10 litres which should see us through.

Π What principles would they have adopted to ensure that the sample was truly representative?

Samples would have been taken from various positions at random across the whole width of the river and at various depths using a sampler of perhaps 250 cm^3 capacity. These would have been combined to give our representative sample.

(*a*) *Preparation of Reagents*

As always, the first task is to assemble the necessary chemicals and make up solutions to the prescriptions given in the paper.

We shall need a litre of good quality chloroform for the preconcentration procedure and 200 cm^3 of cyclohexane for the extraction of the fluorescent species. Both may have to be redistilled in all-glass apparatus before use, though if we use spectroscopically pure cyclohexane this should not require further treatment.

Next, we must make up solutions of the two complexing agents as follows:

(*i*) APDC solution (2% w/v). We dissolve ammonium 1-pyrrolidine dithiocarbamate (10 g) in deionised water and dilute to 500 cm^3. We then purify it by extracting with three 25 cm^3 portions of chloroform.

(*ii*) DAN solution (0.5% w/v). We dissolve 2,3-diaminonaphthalene (0.1 g) and hydroxylamine hydrochloride (0.5 g) in hydrochloric acid (0.1 mol dm^{-3}) and adjust the volume to 200 cm^3. We have to heat this solution for 20 minutes on a water-bath at 50 °C and, after cooling, extract it with three 15 cm^3 portions of cyclohexane. (This solution does not keep well and the authors advise making it up fresh each day.)

Now we have to make up some general purpose solutions of acids and the like by appropriate dilution of AR-grade concentrated solutions as follows:

(*iii*) Hydrochloric acid (6 mol dm^{-3}) (600 cm^3 of conc. acid in one litre)

(*iv*) Hydrochloric acid (1 mol dm^{-3}) (100 cm^3 of conc. acid in one litre)

(*v*) Nitric acid (8 mol dm^{-3}) (500 cm^3 of conc. acid in one litre)

(*vi*) Perchloric acid (70% w/v) (as received)

(*vii*) Ammonia solution (7.5 mol dm^{-3}) (500 cm^3 of 0.880 ammonia in one litre)

(*viii*) Ammonium acetate buffer (ammonium acetate (170 g) dissolved in deionised water and diluted to 500 cm^3)

On this occasion no sample preparation is necessary but we shall have to make up standard solutions of sodium selenite (Na_2SeO_3) and sodium selenate (Na_2SeO_4). The authors specify '1 mM' stock solutions but leave us to work out the quantities involved.

Π How much of each salt shall we need to weigh out to make up 1 litre of stock solution? Can we do this in one step?

[A_r Se = 79.0, Na = 23.0, O = 16.0]

We shall need 0.173 g of sodium selenite and 0.189 g of sodium selenate (1/1000 of the molar mass in each case).

We can weigh out both amounts to sufficient accuracy to enable us to make up 1 litre of solution in a single step.

Finally, we need the tellurium carrier solution for the determination of total selenium. We prepare this by dissolving sodium tellurite (0.868 g) in sufficient hydrochloric acid (1 mol dm^{-3}) to make 500 cm^3 of solution and purify it by adding APDC (2% w/v) solution (5 cm^3) and extracting with three 25 cm^3 portions of chloroform.

We now have all the reagents we need but there are a few items of equipment which we must assemble before we start work. We shall be operating on a 'semi-bucket' scale because we use a 1 litre sample so we must make sure that the larger sizes of laboratory glassware are to hand. In particular we shall need a 1500 cm^3 separating funnel for the chloroform extraction if we can find one, otherwise the 2 litre size will have to do. (The authors specify a 1 litre separating funnel but this does not allow much space for efficient shaking, especially when we have added another 20 cm^3 of reagents and 20 cm^3 of chloroform – in fact, their suggestion does not look to be very sensible!) The smaller sizes of separating funnel required for later stages of the procedure should not present any problem. We shall also need a 2 litre beaker for the total selenium procedure.

We shall need two further items of more specialised equipment:

(*i*) a pH meter and a buffer solution to calibrate the scale near pH 1.

(*ii*) Gelman Grade A/E glass fibre filters about 10 cm in diameter. If the laboratory stores don't carry filters of this type we shall have to make some enquiries as to whether an equivalent grade (ie having similar particle size retention) from another manufacturer is available and, if not, we shall have to order some. Let's assume however that we are in luck and the stores can provide us with a box of 10 cm circles of the correct type of filter together with a Buchner funnel of the right size to take them.

Now we should be ready to start.

(*b*) *Preconcentration Procedures*

If the sample contains suspended particles, it will be advisable to filter it first. If not, we transfer exactly 1 litre to our 1500 cm^3 (or 2 l!) separating funnel and adjust the pH to 4.2 by adding hydrochloric acid (6 mol dm^{-3}, 2.5 cm^3) and ammonium acetate buffer (5.2 cm^3). This is the most favourable pH for the formation of the APDC complex which we produce by adding the APDC reagent (10 cm^3) and mixing thoroughly. Next we add chloroform (20 cm^3) and shake the mixture vigorously for 2 minutes.

When the phases have separated, we run the organic (lower) phase into a 125 cm^3 separating funnel, into which we have previously poured nitric acid (8 mol dm^{-3}, 25 cm^3), and extract the aqueous phase with a further portion of chloroform (15 cm^3). We add this chloroform layer to the first portion in the 125 cm^3 separating funnel and leave it to stand for 8 hours with occasional shaking. (This will probably mean leaving the sample overnight unless we are prepared to work overtime!)

Although this delay is a rather tiresome and unsatisfactory feature of the method, we must not be tempted to take a short cut or we shall fail to recover all the selenium and our result will be low.

Eventually, we shall transfer the aqueous phase to a 50 cm^3 beaker, add perchloric acid (70% w/v, 3 cm^3), and evaporate the solution to near dryness. We can, in fact, usefully occupy the time taken by the back extraction by running 1 litre of distilled water through the entire procedure as a blank. Let's do it ...

We can also carry out the preconcentration procedure for total selenium. Again we take exactly 1 litre of our river water sample (after filtering if necessary) in a 2 litre beaker and add tellurium carrier solution (6 cm^3), hydrazine sulphate (4 g) and concentrated hydrochloric acid (300 cm^3). We must heat the solution to boiling very slowly, boil for 15 minutes with the beaker covered with a clock-glass and then allow it to cool to room temperature. This step is very important because, if we heat the solution too rapidly, we run the risk of producing a precipitate which is too fine to filter.

We must also repeat the procedure up to this point on 1 litre of distilled water as a blank. The precipitate from the blank is processed alongside the sample through the remaining steps of the procedure.

When the solution is at room temperature, we filter it through our glass fibre filter under gentle suction remembering that the filter flask must have a capacity of at least 1 litre if we are to avoid having to interrupt the filtration to empty it.

The next instruction in the paper is:

'Dissolve the precipitate in 10 cm^3 of concentrated nitric acid and 3 cm^3 of 70% perchloric acid'.

Π How, precisely, shall we set about this?

The precipitate must be quantitatively extracted from the filter – simply scraping it off will not be good enough. This will not be all that easy since we are restricted to 13 cm^3 of acid mixture for the dissolution process. Pouring it over the filter (not under vacuum) and collecting the filtrate may not give sufficient contact time to dissolve all the precipitate. The best way is probably to remove the filter from the funnel, fold it so that it can be placed flat in a 100 cm^3 beaker and pour the acid mixture on top of it. It would probably be wise to heat the mixture to near boiling for a few minutes to ensure complete dissolution of the precipitate – we should be able to see when it has all gone because of its dark red colour. When the solution is cool we can remove the filter with acid resistant tweezers (made of teflon or stainless steel) and wash off all traces of the solution back into the 100 cm^3 beaker with a jet of water from a wash bottle. The precipitate from the blank is dissolved up in the same way. Finally, we have to evaporate the solution and the blank down to near dryness (remembering to cover the beakers with clock-glasses to avoid losses due to splashing).

Notice that with both preconcentration procedures we finish up with the sample in identical form – a few cm^3 of acid solution.

Π What is the purpose of this evaporation with perchloric acid?

To destroy all traces of organic matter in the sample which might interfere with the formation of the DAN complex with selenium. Although river water contains a surprising amount of organic material (derived from the peaty soils through which it flows together with contributions from industrial and agricultural sources), the double extraction and co-precipitation procedures should avoid more than a trace of organic material in the final solution. (There may also be a trace of the organic reagent used to extract Se(IV).) This is fortunate because we shall need only a small quantity of perchloric acid. Destruction of organic matter with hot concentrated perchloric acid, though efficient, has a bad reputation because it has occasionally resulted in serious explosions – in fact, some laboratories ban its use altogether. However, the nitric/perchloric acid mixture is far safer since the bulk of the degradation of the organic material is accomplished in the presence of the nitric acid before the concentration of the perchloric acid rises to a maximum in the final stages to deal with the last persistent traces.

(*c*) *Calibration*

Another aspect of the method which is rather vague in this paper is the calibration procedure. No actual calibration graph is presented and we appear to be using the 'method of standard additions' though the authors do not actually say so! We have to add 'known amounts of selenium(IV) to 1 litre aliquots of the sample' which implies that a separate calibration is carried out for each sample. In practice, it may well be possible to use a single calibration for all samples after the first where all the samples are very similar. Fortunately we have only one sample - though we shall want to run a duplicate.

We can get a clue as to the amount of selenium to add to give measurable fluorescence from the data given in the paper for the experiments on the recovery using these procedures. The range of concentration of selenium used for the APDC extraction was 0.1–4.8 nmol dm^{-3} while in the case of coprecipitation with tellurium it was 0.1–0.4 nmol dm^{-3}. A concentration of 1.0 nmol dm^{-3} would therefore seem reasonable.

Π How much of our 1.0 mmol dm^{-3} sodium selenite stock solution shall we have to add to one litre to get an 'added selenium' concentration of 1.0 nmol dm^{-3}?

1.0 μl. We have to dilute by a factor of 10^6.

Π There are two problems in making up a solution by adding 1.0 μl of stock solution to one litre of sample. What are they and how can we minimise them?

(*i*) It is very difficult to dispense 1.0 μl of liquid accurately under routine laboratory conditions.

(*ii*) Even with modern micro-pipettes errors of up to 0.2 μl are possible so a volume of 10 μl would be required to give an accuracy better than $\pm 2\%$.

It is difficult to ensure complete mixing of two solutions whose volumes differ by a factor of a million.

To minimise these problems it would be better to dilute the stock solution to 1.0 μmol dm^{-3} so that we can add 1 cm^3. If we have a modern micro-pipette we can do this with an accuracy of better than $\pm 1\%$ by diluting 100 μl to 100 ml. Otherwise we shall have to do this in two steps, 5 ml$\rightarrow$100 ml followed by 2 ml$\rightarrow$100 ml. Let's carry out this dilution to give us a 1.0 μmol dm^{-3} standard.

Since the selenium level is entirely unknown to us, as a first shot we might well run a sample with 1 nmol dm^{-3} added selenium alongside the samples we have already processed. The addition of 1 cm^3 of our 1.0 μmol dm^{-3} standard to a litre increases the volume by 0.1% which is negligible compared with the overall precision of the method.

It may have occurred to you that it would be quicker and altogether more convenient to make standard additions to the sample after the preconcentration stage when we have only a small volume to deal with.

Π Why do you think that the authors stipulate adding the standard to the 1 litre sample at the start of the procedure?

It is more accurate since it allows for possible loss of selenium during the preconcentration procedures. The authors found that recovery of selenium was only about 90% at best so their suggestion is sound.

Since the coprecipitation procedure is much quicker than the chloroform extraction we can use that to determine whether we are adding a reasonable amount of selenium. Let's take another 1 litre of sample, add 1 cm^3 of our sodium selenite standard (1.0 μmol dm^{-3}) solution and put it through the precipitation procedure – then we can go home!

(d) Measurement of Fluorescence

We now return to the laboratory the morning after carrying out the preconcentration procedures. The back-extraction of the chloroform extract of selenium(IV) will be complete and so, after a final shake, we can discharge the chloroform (layer) into the waste bottle (not down the sink!) and run off the aqueous layer into a 50 cm^3 beaker. After adding perchloric acid (3 cm^3) we evaporate the solution to near dryness taking the usual precautions to prevent splashing. Of course, we have to complete the processing of the blank in the same way.

We then have three solutions ready for fluorescence measurement – this one and the two prepared by the coprecipitation technique, one having received the standard addition. There will also be the two blanks, one for each procedure.

Before we do anything else, it would be a good idea to switch on the spectrofluorimeter so that it will have settled down by the time we come to read the fluorescence. Now for the solutions.

To each of our preconcentrated samples and the two blanks in their 50 cm^3 beakers we add hydrochloric acid (6 mol dm^{-3}, 10 cm^3) and boil for 5 minutes (with lids on). After cooling we adjust the pH to 1.0 with ammonia solution (7.5 mol dm^{-3}), testing with the pH meter. Next we add the disodium salt of EDTA (0.1 mol dm^{-3}, 0.5 cm^3) and the DAN reagent solution (10 cm^3) Then we transfer the solutions quantitatively to 50 cm^3 standard flasks, washing out the beakers and making up to the mark with hydrochloric acid (0.1 mol dm^{-3}). To ensure complete reaction we heat the solutions to 50 °C for 20 minutes and then allow them to cool.

We shall have to extract each of these solutions with cyclohexane (5 cm^3). The authors advise centrifuging the organic phase at 2500 rpm for 15 minutes before measurement which suggests they had some problem with separation of the phases. We may have to resort to this if the organic phase is cloudy but, with any luck we may get away without it. Our sample is fresh water which may make all the difference.

We should have just enough cyclohexane extract to take a duplicate reading but we cannot use the triple wash procedure on this occasion (unless we a low volume cell of some type to minimise sample volume). With a standard 1 cm cell we can only empty the cell and drain it as far as possible.

Before we process the next sample we shall have to wash out the separating funnel first with acetone and then with cyclohexane.

Π Why do we need to employ these two solvents in the washing procedure?

Acetone is miscible with both the organic and aqueous phases and is used to remove all trace of the last sample. Cyclohexane is used to remove the acetone and any fluorescent impurities it may contain.

Π The cell will also have to be washed out with cyclohexane and dried – how shall we dry it and why?

The safest method is to draw a stream of air through the cell using a filter pump. Blowing air from a compressor is not recommended because it usually contains droplets of oil which could play havoc with fluorescence measurements. Don't attempt to dry it in an oven unless you are *sure* it is safe with organic vapours. A spark from the ordinary type of thermostat control could cause an explosion! Drying is necessary because we do not have enough sample to wash out the cyclohexane which would otherwise dilute the next sample.

Π Before we measure the three solutions and the two blanks, which do you think will give the highest reading and what order will the others fall in? Place the five solutions in the likely order of *decreasing* fluorescence intensity (ie highest first).

Precipitation sample + standard > precipitation standard > chloroform-extracted sample > precipitation blank > chloroform-extracted blank.

The coprecipitation method measures the total selenium content and so will probably give a higher reading than the chloroform extraction which is specific for selenium(IV). [There may, of course, be no Se(VI)]. Obviously the sample to which we added the selenium standard will give the highest reading. The two blanks could go either way, but it is perhaps more likely that precipitation blank will be the higher.

Π Which reagent is the most likely to be the reason for this?

The sodium tellurite solution which could well contain traces of selenium.

Well, we have thought about what we are going to do (which is always good practice in any analysis!) so let's get on and do it.

We transfer the first of these solutions (precipitation sample + standard) quantitatively to a clean 125 cm^3 separating funnel (rinsing with hydrochloric acid (0.01 mol dm^{-3}), add cyclohexane (5 cm^3), shake the mixture vigorously for 2 minutes and wait for the phases to separate. Yes, we are in luck! The phases have separated cleanly and the organic phase is quite clear.

Π Which layer *is* the organic phase?

On this occasion the organic phase is the *upper* layer – hydrocarbons are invariably lighter than water. We therefore run off the lower layer and discard it, run the organic phase into our fluorescence cell and measure the fluorescence at 520 nm with excitation at 380 nm. No band width is specified so let's try 10 nm for both monochromators. We get a value of 522 for this solution which is reasonable for what we expect to be the highest reading, giving us three figure precision for the samples and providing an adequate margin in case our predictions are wrong.

We now wash out the cell and separating funnel (in that order so that the cell will go on drying while we wash the funnel) with acetone and cyclohexane as we previously planned and proceed to measure the other four solutions in the same way as the first. We obtain the following set of readings:

Precipitation sample + standard	522
Precipitation sample	384
Chloroform-extracted sample	347
Precipitation blank	28
Chloroform-extracted blank	12

We can use these readings to get a rough estimate of the concentrations of Se(IV) and the Se(VI) in the river-water sample based on a single standard addition and the assumption that the calibration is linear and identical for both methods.

Π Have a shot at working these values out. Does it look as if we might have a pollution problem?

Total selenium $= 0.204\ \mu g\ dm^{-3}$
Selenium (IV) $= 0.192\ \mu g\ dm^{-3}$
Selenium (VI) $= 0.012\ \mu g\ dm^{-3}$

These values are well below the toxic limit ($1.0\ \mu g\ dm^{-3}$) so we do not appear to have a pollution problem.

Let's see how these values are derived. First subtract the appropriate blank value from the solution values:

Precipitation sample + standard $= 522 - 28 = 494$
Precipitation sample $= 384 - 28 = 356$
Chloroform-extracted sample $= 347 - 12 = 335$

The increase in concentration due to the addition of $1\ cm^3$ of $1\ \mu mol\ dm^{-3}$ standard to 1 litre is $1.0\ nmol\ dm^{-3}$.

This causes an increase in the fluorescence reading of

$$494 - 356 = 138$$

If we assume that the fluorescence versus concentration graph is linear over this region the concentrations of the two samples with no addition are:

Precipitation sample

$$= 356/138 \times 1.0\ nmol\ dm^{-3}$$

$$= 2.58\ nmol\ dm^{-3}$$

$$= 2.58 \times 79.0 = 204\ ng\ dm^{-3}$$

Chloroform-extracted sample

$$= 335/138 \times 1.0\ nmol\ dm^{-3}$$

$$= 2.43\ nmol\ dm^{-3}$$

$$= 2.43 \times 79.0 = 192\ ng\ dm^{-3}$$

The former refers to the total selenium and the latter to Se(IV). The difference, 12 ng dm^{-3}, therefore corresponds to Se(VI). This is only about 6% of the other values and so may not be significant because we have taken a small difference between two large numbers whose precision is only ± 5%.

Π On the basis of these results suggest a standard additions procedure for refining them. (Remember that we have already used 3 litres of our sample.)

Add 1, 2 and 3 cm^3 of both 1 μmol dm^{-3} standards to three 1 litre aliquots for the precipitation procedure.

Add 2, 4 and 6 cm^3 of the selenite standard and 1, 2 and 3 cm^3 of the selenate standard to three further 1 litre aliquots for the chloroform extraction procedure.

Π We should then get similar curves for the two methods – why?

The precipitation method measures total selenium whereas the chloroform extraction method measures only Se(IV). The second batch of aliquots each contain as much selenium as Se(IV) as the total selenium in the corresponding aliquot in the first batch.

The selenium added is thus approximately 1, 2 and 3 times the quantity already present.

Π What is the basis of this particular choice of concentrations of added selenium?

It will give a slope of about 45° to the calibration graph which should provide the most accurate extrapolation to the concentration axis.

Let's now take 6 one litre aliquots, make the appropriate standard additions to each group of three, and carry them through the two procedures. We don't have enough sample to run duplicates but the solutions are related to each other in a systematic way so we can estimate our error statistically by the spread of the results about the best straight line through the experimental points (or smooth curve if that gives a better fit). We can, of course, include our previous preliminary results in the data which will provide the blank and zero addition values. We also have about a litre of sample left over in case of accidents!

When we have done this we get the following fluorescence readings corresponding to the six aliquots to which the standard additions were made:

Precipitation sample +	
1 cm^3 addition of each standard	670
2 cm^3 addition of each standard	928
3 cm^3 addition of each standard	1212
Chloroform-extracted sample +	
2 cm^3 of Se(IV) standard	608
4 cm^3 of Se(IV) standard	898
6 cm^3 of Se(IV) standard	1166

Π Use these readings together with those given in the previous table to obtain concentrations of Se(IV) and Se(VI) in the river water sample.

Total selenium = 0.202 μg dm^{-3}

Selenium(IV) = 0.192 μg dm^{-3}

Selenium(VI) = 0.010 μg dm^{-3}

As before, we first subtract the appropriate blank value (taken from the previous data) from each of the solutions and plot these values against the added concentration of selenium. Remember that the addition of 1 cm^3 of the 1.0 μmol dm^{-3} standard increases the concentration by 1.0 nmol dm^{-3}. Draw the best straight lines through both sets of data (for the precipitation sample and the chloroform-extracted sample) and extrapolate back to the concentration axis. Check whether a smooth curve would fit the data

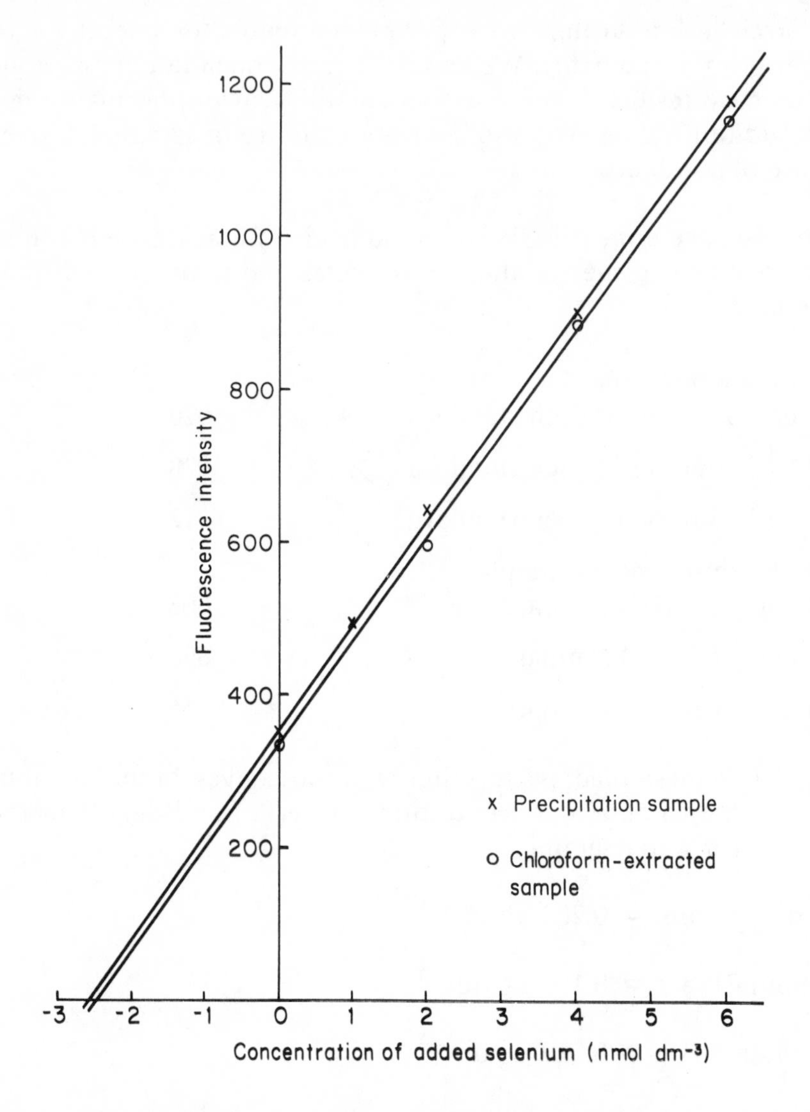

Fig. 5.6a. *Calibration graphs for the determination of selenium by a standard addition method*

better – it wouldn't in either case. Read off the (negative) intercept on the concentration axis:

Precipitation sample (Total selenium)

= 2.56 nmol dm^{-3}

= 202 ng dm^{-3}

Chloroform-extracted sample (Selenium(IV))

= 2.46 nmol dm^{-3}

= 192 ng dm^{-3}

The actual graphs plotted from these data are shown in Fig. 5.7a.

The values obtained from the standard addition procedure are not significantly different from our first estimate but the Se(VI) concentration value is now supported by the general trend of the experimental values. However, it is very small and probably not important anyway.

Π How do you account for the good agreement between our first estimate and the refined standard addition values?

The assumptions that we made about the linearity of the calibration and the similarity between the samples were justified.

The cynical answer is perhaps that the results are concocted and not subject to the influences present in real samples! Even so the point made is still valid.

5.7. FLUOROIMMUNOASSAY OF PROGESTERONE IN HUMAN SERUM

For our example of FIA we shall study a straightforward application of the method to the determination of a female sex hormone, progesterone, in blood serum.

We are also going to revert to using l instead of dm^3 and ml instead of cm^3. You have to be familiar with both and particularly the former when reading the clinical chemistry literature.

The concentration of progesterone in the blood serum of women varies in a systematic way during the menstrual cycle from about 1 ng ml^{-1} in the first half of the cycle and rising to a maximum of about 10 ng ml^{-1} after ovulation (Fig. 5.7a). During pregnancy the level rises dramatically to a maximum of about 160 ng ml^{-1} in the late stages falling rapidly during the last two weeks to normal levels (Fig. 5.7b). Clinical interest in progesterone is directed mainly towards its use as an indicator that ovulation has occurred which is of fundamental importance in studies of fertility. Although it is the most abundant of the hormones associated with pregnancy, the levels of other hormones are much more useful as a guide to whether pregnancy has occurred and in monitoring its subsequent progress. The paper from which the method is taken is quite recent (Allman,

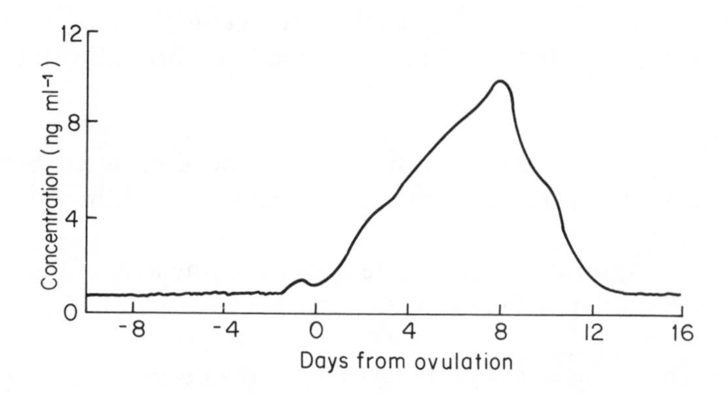

Fig. 5.7a. *Progesterone levels during the menstrual cycle*

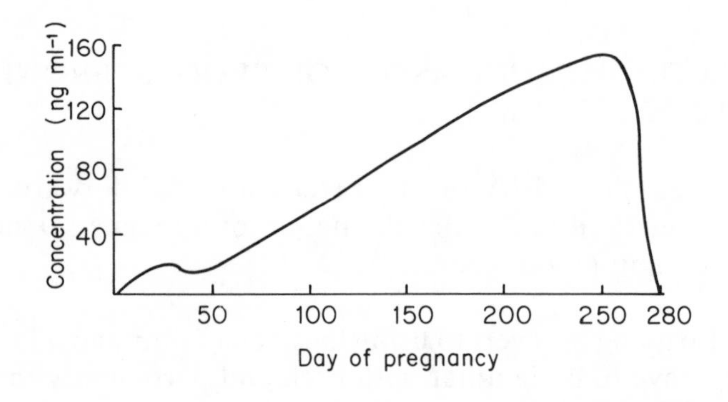

Fig. 5.7b. *Progesterone levels during pregnancy*

Short, and James, *Clinical Chemistry*, 1981, **27**, 1176) and is of particular interest because the results of FIA are compared directly with those of RIA on the same sample. A check was also made on the precision and recovery of the method.

5.7.1. Basis of the Method

Progesterone in the form progesterone-3-carboxymethyloxime is labelled with fluoresceinamine.

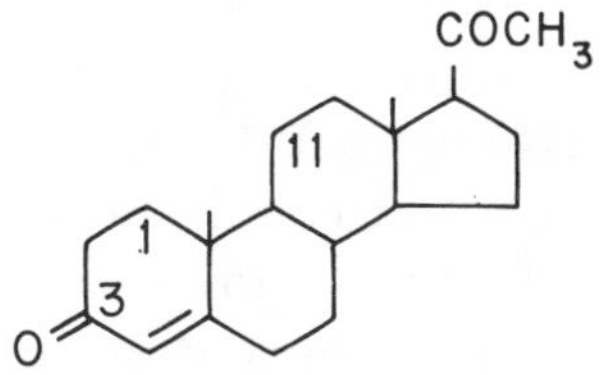

COCH$_3$ HO O OH O C=O NH$_2$

Progesterone 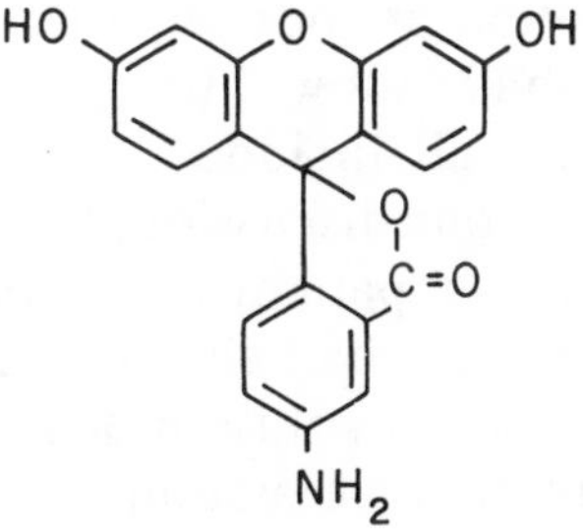 Fluoresceinamine

An antiserum, raised in a goat, against progesterone-11-hemisuccinate linked to bovine albumin is used as the antibody. The progesterone is extracted from samples of blood serum with hexane. The hexane extract and a set of progesterone standards are pipetted into incubation tubes and evaporated to dryness in a vacuum oven at 40 °C. 300 μl of an assay solution containing plasma albumin, gamma globulin, labelled progesterone and antiserum in appropriate amounts in a borate buffer solution at pH 8.0 is added to each sample and standard mixed thoroughly and incubated overnight at 4 °C. The gamma globulins are then precipitated with ammonium sulphate carrying down the progesterone/antibody complex. This is separated completely from the supernatant liquid and dissolved in sodium hydroxide solution (0.1 mol l^{-1}). The fluorescence of this solution is measured at 530 nm with excitation at 490 nm. A calibration graph of fluorescence intensity versus concentration is plotted using the values for the standards from which the concentration of the samples can be determined.

Π Would this assay be described as a homogeneous or heterogeneous assay?

It is heterogeneous. The bound antigen is physically separated from the free antigen.

As an alternative to liquid phase incubation, the authors investigated an ingenious method in which the antiserum is added in the solid phase, covalently linked to magnetisable cellulose particles to allow the antibody/antigen complex to be separated by magnetic sedimentation. The time for the analysis is also reduced by carrying out the incubation at room temperature with constant mixing for one hour. We shall follow through the liquid phase method in this study though we shall include the authors results with the solid phase technique in our discussion. We shall assume that the blood sample taken from our patient has already been pretreated and comes to us as serum. We shall also assume that the antiserum is available in our laboratory in solution at an appropriate concentration. These are both reasonable assumptions in the context of an analysis in a hospital laboratory since pretreatment of blood samples is a routine operation applied to 'bulk' samples before the serum is distributed to different operators responsible for particular analyses. The antiserum would also have been purchased from a commercial supplier probably as part of a 'kit' for carrying out this particular assay. The kit would also include the labelled antigen (as well as most of the other reagents required) but, since this is of direct relevance to the fluorescence measurement, we shall carry out this operation ourselves.

5.7.2. Laboratory Procedure

(a) Preparation of the Labelled Antigen

The label is basically fluorescein though we shall actually use it as its amine derivative rather than the more common isothiocyanate. The reaction is quite complex so we shall just follow the recipe without worrying about the organic chemistry concerned – the authors are distinctly unhelpful on this point! We shall require about 10 mg each of

(*i*) Progesterone-3-carboxymethyloxime
(*ii*) Fluoresceinamine

These two solid reagents must be dried under vacuum over silica gel. Fluoresceinamine is relatively cheap (10 pounds sterling per gram) whereas the progesterone derivative is very expensive (1 pound per mg) so we shall have to purchase the minimum quantity for our needs. (The labelled derivative would cost about 10 pound per mg so it will be a worthwhile saving to make it ourselves). We shall also need

(*iii*) Isobutyl chloroformate
(*iv*) N-methylmorpholine

These two liquid reagents have to be dried over molecular sieve. We only need a few microlitres of each but as they are quite cheap it will be easier to dry them in bulk. To about 5 ml of each reagent we shall add molecular sieve (type 4A, 1/16-inch pellets – about 5 g) and leave them to stand overnight in a well-stoppered glass vial. We shall also dry the solvent, dimethylformamide, in the same way – again about 5 ml will be sufficient. Like most organic solvents, dimethylformamide is volatile and fairly toxic so we must handle it under a fume hood. We shall also need to redistil about 20 ml of hexane and 200 ml of ethanol for use later as solvents. When our reagents are dry the first step is to dilute the two liquids 1 : 10 with dimethylformamide. We take 0.1 ml of each reagent in separate glass vials (again with well fitting stoppers to keep them dry) and add 0.9 ml of dimethylformamide to each.

Π After mixing, the solutions must be cooled to −5 °C. How shall we do this?

A simple ice and salt freezing mixture will be quite suitable – provided the laboratory has a supply of ice.

Now we take 5 mg of progesterone-3-carboxymethyloxime, dissolve it in dimethylformamide (50 μl) and again cool the solution to −5 °C. Then we add the isobutyl chloroformate solution (15 μl) followed by the N-methylmorpholine solution (12.5 μl) and leave the solution at −5 °C for the reaction to proceed. While this is happening, we dissolve fluoresceinamine (4 mg) in dimethylformamide (50 μl) and cool the solution to -5 °C. When the reaction has proceeded for 35 minutes we pour in the whole of the fluoresceinamine solution and let the reaction proceed for a further 1 hour at -5 °C and then leave it overnight at 7 °C – well-stoppered of course.

The labelled antigen has to be purified by thin layer chromatography on silica gel. Commercial sheets are readily available and we shall need 12 sheets of a grade suitable for preparative work with no fluorescent indicator. The solution is diluted with redistilled ethanol (2 ml) and applied to 12 tlc sheets in streaks – the usual preparative procedure. The chromatogram is developed with a solvent system consisting of chloroform/ethanol/water (87/12/1 by volume) in the usual way and the areas of the sheets containing the reaction product cut out.

Π How shall we know where the material is on the sheet?

We have just prepared a fluorescent derivative of progesterone so we should have no difficulty in locating it by looking at the sheet under a uv lamp – hence the reason for specifying tlc sheets with no fluorescent indicator. The excess fluoresceinamine will probably fluoresce too but we may reasonably assume that it will have moved farther from the starting zone than the much larger labelled progesterone derivative.

When we have cut out the required portions of the sheets we extract the reaction product with redistilled ethanol (total volume 50 ml) and store the solution in the dark at 4 –7 °C until required. We shall have to determine the concentration of the progesterone–fluorescein derivative before we use it by measuring the absorbance at 491 nm in sodium hydroxide solution (0.01 mol dm^{-3}). We shall assume that the absorptivity is the same as for fluorescein itself, 88 000 l mol^{-1} cm^{-1}.

(*b*) *Preparation of Standards*

We need to calibrate the serum over the concentration range 0–100 nmol l^{-1} with seven standards and a blank. The sample solution is obtained by extracting serum (200 μl) with redistilled hexane (2 ml) and 500 μl is used in the assay. The standards are dissolved in redistilled ethanol and 100 μl is used in the assay.

Π How shall we make up the standard which will be equivalent to 100 nmol l^{-1} in the serum? (Relative molar mass of progesterone = 312. Unlike its 3-carboxymethyloxime, progesterone itself is relatively inexpensive (1 pound per gram) so we do not need to use micro-techniques to prepare the standards even though we shall use only 100 μl in the assay.)

Dissolve progesterone (156 mg) in redistilled ethanol and make up to 100 ml, then dilute 1 ml of this stock solution to 100 ml. Finally take 1 ml of the diluted solution and dilute it further to 100 ml. This gives us a primary standard containing 156 ng ml^{-1}. With good analytical technique the solution should be accurate to $\pm 1\%$ which is more than adequate in this type of analysis. With modern microsyringes in the hands of a proficient operator, the 100-fold dilution can be done by transferring 0.1 ml into 10 ml without loss of precision and with a saving in the redistilled solvent.

10-fold dilution of the primary standard (1 ml→10 ml) gives a working standard containing 15.6 ng ml^{-1} so that the 100 μl taken for assay contains 1.56 ng of progesterone. Since the relative molar mass of progesterone is 312, 1.56 ng corresponds to 1.56/312 nmol = 0.005 nmol = 5 pmol.

If serum contains progesterone at a concentration of 100 nmol l^{-1}, the amount of progesterone in the 200 μl we take for assay is

$$[200 \times 10^{-6} \times 100 \text{ nmol} = 20\,000 \times 10^{-6} \text{ nmol}]$$

$$[= 20 \times 10^{-3} \text{ nmol} = 20 \text{ pmol.}]$$

This is extracted into 2 ml of hexane so the 500 μl of the extract used in the assay contains 5 pmol of progesterone.

To make up the other six working standards we can now take 0.8, 0.6, 0.4, 0.2, 0.1, and 0.05 ml of the primary standard and dilute each to 10 ml.

Π What will be the serum concentration corresponding to these concentrations?

80, 60, 40, 20, 10, and 5 nmol l^{-1} since a dilution of 1 ml to 10 ml gives a standard corresponding to 100 nmol l^{-1}.

(*c*) *Preparation of the Assay Buffer*

We must also prepare a borate buffer solution by dissolving AR grade boric acid (0.31 g) in deionised water (80 ml), adding sodium hydroxide solution (1.0 mol l^{-1}) to adjust to pH 8. This buffer solution forms the basis for a composite assay solution which contains the labelled antigen and the antibody as well as two proteins, albumin and gamma globulin. Now is a convenient time to add these components to the buffer which we shall then make up to 100 ml. The proteins are no problem – we simply add plasma albumin (0.1 g) and gamma globulin (0.1 g) and shake the solution thoroughly until they are dissolved. We now have to add 0.1 μg of our labelled progesterone so we must first determine the concentration of the stock solution we prepared earlier. This is done by measuring the absorbance at 491 nm in sodium hydroxide solution (0.01 mol l^{-1}) and using the absorptivity (88 000 l mol^{-1} cm^{-1}). Our progesterone-fluorescein solution is in ethanol so we must add 1 ml to the sodium hydroxide (0.01 mol l^{-1}) in a 10 ml standard flask and make up to the mark with sodium hydroxide (0.01 mol l^{-1}).

Suppose we do this using 5 mm cells and get an absorbance reading of 0.814. That corresponds to a concentration of our progesterone–fluorescein stock solution of 120 μg l^{-1}, by a calculation as follows:

applying the Beer–Lambert Law $A = \epsilon c d$

$$0.814 = 88\,000 \times 0.5 \times c$$

$$= 44\,000 \times c$$

$$c = 0.814/44\,000$$

$$= 0.0185 \times 10^{-3} \text{ mol l}^{-1}$$

Assuming the labelled compound is 1 : 1 progesterone : fluorescein we can assume a relative molecular mass of 650 to a close enough precision. Hence the concentration of the solution we measured is

$$\begin{aligned} 0.0185 \times 650 \times 10^{-3}\ \text{g l}^{-1} &= 0.0185 \times 650\ \text{mg l}^{-1} \\ &= 0.0185 \times 650\ \mu\text{g ml}^{-1} \\ &= 12.0\ \mu\text{g ml}^{-1}. \end{aligned}$$

The solution was obtained by diluting the stock solution $\times 10$ with sodium hydroxide so the concentration of the stock solution is 120 μg ml^{-1}.

Π We have to add 0.1 μg to our 100 ml of assay buffer. How shall we do this? Do we have to worry about the precision of this quantity?

We can simply add 1 μl of the stock solution to the assay buffer since this volume contains 0.12 μg of the progesterone–fluorescein complex. This will require a 1 μl microsyringe and, provided you are experienced in the dispensing of these small volumes (which is a routine operation in gc and hplc techniques), adequate precision will be obtained. Alternatively we could add 10 μl of the solution we used for the absorbance measurement if you were not confident in handling the smaller volume. (The addition of 10 μl of sodium hydroxide solution (0.01 mol l^{-1}) will not affect the pH of the buffer significantly). The precision of this measurement is not very important and the fact that we are adding 0.12 μg rather than 0.10 μg is of no consequence. The important thing is that the amount is constant for each determination since the method is calibrated with standard solutions and the same volume of assay buffer is used in each case. Of course, if we were calculating our results from the equations we should have to know the concentration of labelled antigen to a precision consistent with the other values and which would be reflected in the precision of our results.

Finally we have to add the antiserum to the buffer. Again the absolute quantity is not important provided it is of the right order to ensure that the total antigen is in excess, and equally important, constant in every determination. In most cases the antiserum is calibrated by the manufacturer and all we have to do is dilute it by an appropriate amount.

Π The authors of this method specify a final dilution of 3000 for the batch of antiserum they used so let's assume we have a sample from the same batch. How much do we add to the assay buffer?

33.3 μl would give a dilution factor of exactly 3000. In practice, provided we add between 30 and 35 μl (from a calibrated 100 μl syringe), we shall be near enough.

We have now added all the components necessary for the assay to the buffer and we can make it up to 100 ml.

(*d*) *Assay Procedure*

We can assume that we have received some samples of serum for analysis. The first step is to extract the progesterone into redistilled hexane which we do by vigorously shaking 200 μl of the serum with 2 ml of hexane for 15 minutes. We shall want some mechanical assistance here – the authors specify a multi-vortex shaker to ensure efficient mixing (a small scale version of a food liquidiser with several rotors). 500 μl of each sample extract is then transferred to a separate incubation tube. We then pipette 100 μl of each standard solution into a similar set of incubation tubes and place these and the tubes containing the samples into a vacuum oven and then evaporate to dryness under reduced pressure at 40 °C.

When the evaporation is complete we remove the tubes from the oven and, when they have cooled to room temperature we add 300 μl of the assay buffer to each, mix them thoroughly with our vortex mixer, and incubate them overnight at 4 °C.

We start the new day by making up a saturated solution of ammonium sulphate solution, cooling it to 4 °C and adding 300 μl to each tube. This precipitates the gamma-globulin which we centrifuge so that we can decant the supernatant liquid. It is important to remove all traces of liquid adhering to the tube and to the gamma-globulin which forms a 'pellet' during centrifuging. This is done by using a Pasteur pipette connected to a vacuum line. Finally we add sodium hydroxide solution (0.1 mol l^{-1}, 650 μl) to dissolve the globulin pellet.

We now have to measure the fluorescence of these solutions so we switch on the spectrofluorimeter and set the excitation wavelength to 490 nm and the excitation wavelength to 530 nm. The authors used slit widths of 10 nm for excitation and 20 nm for emission on their Perkin–Elmer Model 3000 fluorescence spectrometer so we must choose similar values on our instrument. (They are usually available on most instruments.) We shall need a low volume cell to hold our samples (650 μl will not go far in a standard 1 cm cell) – the authors used a 5 mm^2 square cell so again we must find something similar though, if no other alternative was available, we could use a microcell of much lower capacity (say, 50 μl).

Π You should be familiar now with the precautions to be taken when measuring fluorescence so it should be safe to leave you to take the measurements! Which solution will you measure first?

The blank. This will contain the most labelled antigen and so give the highest fluorescence reading. You need to read the highest value first so that you can set the sensitivity or scale expansion factor (depending on the type of instrument) to give a reading near the top end of the read-out display (80–90% full scale deflection, whatever form that takes).

The other standards and the samples can then be read off and the readings recorded carefully in your laboratory notebook – unless you are fortunate enough to have a modern instrument which prints out the results for you. (You will still need to keep a careful record of the sample identities however.)

Π To complete the assay you will have to plot a calibration graph using the results from the standards and then read off the concentration of the samples directly. Some modern instruments have an autoconcentration facility but you will not be able to use this for FIA. Why?

The autoconcentration facility assumes that the calibration graph is linear and that the reading increases with concentration, neither of which applies to FIA calibrations. If your instrument has some very sophisticated software associated with it you may find an FIA procedure. However, we shall assume that you will have to draw the calibration by hand.

You may be surprised by the shape of the calibration curve and the fact that the blank gives the highest reading so let's look at the procedure point by point and identify what actually happens.

Π Where does the antibody/antigen reaction take place?

In the incubation tube when it is left to stand overnight.

Π Where does the unlabelled antigen come from?

In the case of samples, from the hexane extract of the serum. In the case of standards it is pipetted directly into the incubation tube.

Π Where does the labelled antigen come from?

It is present in the 300 μl of assay buffer added to the tubes after evaporation of the solvent in the vacuum oven.

Π How is the antibody added to the system?

It is present in the assay buffer as a conjugate with albumin and complexed with the labelled antigen which is in excess.

Π What is the purpose of the extra protein (albumin and globulin) added to the assay buffer?

The exact mechanism of the immune reaction is complicated but we can simply regard the protein as a carrier for the antibody/antigen complex which assists in the separation of the bound from the free antigen.

Π Does the fluorescence measurement come from the bound or free antigen?

From the bound antigen – the free antigen was removed when we decanted the supernatant liquid. In fact, when we measure the fluorescence it actually comes from the free fluorescein label. Dissolving the globulin in fairly concentrated sodium hydroxide hydrolyses the antibody/antigen complex and then hydrolyses the progesterone-fluorescein derivative. This procedure was adopted by the authors because it gave better sensitivity – presumably the fluorescence efficiency of the derivative is less than that of the derivatising agent itself.

The authors give some comparative results using a solid phase assay in which the antiserum is incorporated into magnetisable cellulose particles. This greatly facilitates the separation of the bound antigen requiring only sedimentation on a ferrite magnet which is both more convenient and far less expensive than a centrifuge. The precision is better than that of the liquid phase assay (as might be expected) and compares favourably with RIA.

The FIA method (both versions) shows the expected advantages over RIA associated with the use of a cheap, stable, non-radioactive antigen and relatively inexpensive equipment for measuring the fluorescence. Its chief disadvantages are that it consumes more of the expensive antiserum than RIA (which requires a 25 000-fold dilution compared with 3000-fold in FIA) and, at present at least, measurements have to be made manually taking about 2 minutes per sample. Work is already in hand to automate the process however and the latter disadvantage will undoubtedly disappear as the technique develops.

SAQ 5.7a

Identify in the following list, the operations which have to be carried out precisely (ie to $\pm 1\%$ or better) during the progesterone immunoassay because they affect the precision of the final results:

(*i*) Weighing 156 mg of progesterone to make up the stock solution.

(*ii*) Weighing 5 mg of progesterone-3-carboxymethyloxime to prepare the labelled antigen. ⟶

SAQ 5.7a (cont.)

(*iii*) Measuring the absorbance of the stock solution of labelled antigen.

(*iv*) Measuring the fluorescence of the incubated solutions.

(*v*) Measuring the volume of antiserum to be added to the assay buffer.

(*vi*) Dispensing 100 μl of standard to the incubation tube.

(*vii*) Measuring 2 ml of hexane with which to extract the serum.

(*viii*) Addition of 300 μl of assay buffer to the sample and standards.

SAQ 5.7b

Seven standard solutions, prepared and measured in a fluoroimmunoassay of progesterone, gave the following fluorescence intensities:

Concentration in Serum (nmol l^{-1})	Fluorescence Intensity (arbitrary units)	
0	700	711
5	671	680
10	592	611
20	493	503
40	410	422
60	333	342
80	299	311
100	279	288

The background fluorescence from the reagents other than the progesterone–fluorescein complex is 36 arbitrary units. Plot the calibration graph based on these data.

What is the most striking feature of the shape of this graph? Estimate the dynamic range of the analysis.

SAQ 5.7c

The serum taken from four women patients at 3-day intervals over a period of 2 weeks was analysed by this procedure and gave the following results:

Patient	Day 1	Day 4	Day 7	Day 10	Day 13
A	655	650	548	450	410
B	641	628	637	644	635
C	445	472	642	654	650
D	295	313	335	356	368

Determine the concentration of progesterone in each sample using the calibration graph you have just drawn and give a clinical diagnosis of the condition of the four women (use Figs. 5.7a and 5.7b). In which case would further tests be desirable?

5.8. ROOM TEMPERATURE PHOSPHORESCENCE OF POLYNUCLEAR AROMATIC HYDROCARBONS IN CYCLODEXTRINS

Unlike the examples of analytical methods we have studied so far, the paper which forms the basis for this final case study (Scypinski and Cline Love, *Anal. Chem..* 1984, **56**, 322) does not present a fully worked out procedure for carrying out a specific analysis. The authors have been more concerned with the fundamental processes associated with the enhancement of phosphorescence by the inclusion of the analyte molecules into the cavity in the cyclodextrin structure. However, they have also attempted to assess the analytical potential of the method by establishing the limits of detection and the linear calibration range for one or two typical compounds.

We have in fact come into the story of the development of this technique at a relatively early stage after the initial research observations have been made but before the final development of well-tested recipes which can be adopted as routine laboratory methods. This process can take many years to complete and, of course, all the established analytical methods in common use have been through it. It is therefore entirely appropriate that we should pick up one method at this stage in a study in analytical chemistry.

5.8.1. Basis of the Method

Π The first observations of RTP from molecules trapped in the cyclodextrin cavity were made with the halonaphthalenes (Turro et al, *Photochem. Photobiol.*, 1982, **35**, 69 and 1983, **37**, 149). In this application the halogen contributes a 'heavy-atom effect' which promotes inter-system crossing into the triplet state. The excited molecule is protected from deactivation by the cyclodextrin environment until phosphorescence can occur. The present authors have extended the range of application of the technique by introducing the heavy atom in a separate molecule, 1,2-dibromoethane $BrCH_2CH_2Br$. What advantage does this have over Turro's work?

Turro's method is restricted to compounds which already contain a halogen atom. The inclusion of the heavy atom in a separate molecule makes it possible to excite phosphorescence from molecules containing only carbon, hydrogen, oxygen and nitrogen – in the present case from hydrocarbons.

The luminescent species is described as a ternary complex between the analyte, dibromoethane and cyclodextrin formed inside the 'soft centre' of the cyclodextrin 'doughnut'. The equilibria involved are:

CD
DBE ⇌ K_2 K_1 ⇌ AN
CD:DBE CD:AN
AN ⇌ K_3 K_4 ⇌ DBE
CD:DBE:AN

AN = analyte molecule
DBE = dibromoethane
CD = cyclodextrin
K_1 - K_4 = equilibrium constants

In order for phosphorescence to occur either K_3 or K_4 must be larger than K_1 and K_2. If K_1 is greater than K_4 enhanced fluorescence emission is observed but no phosphorescence.

Π What happens if K_2 is much larger than K_3?

Little fluorescence or phosphorescence is observed because the analyte is not involved in the complex formation.

The relative values of these equilibrium constants depend on the matrix of the analyte, the halogen-containing compound and the cyclodextrin. In this work, dibromoethane was the only halogen compound used. Three different cyclodextrins, α, β and γ which contain 5, 6 and 7 glucose rings respectively were studied. The different sizes of the cavities in these compounds is crucial and, as we shall see, choice of a particular cyclodextrin is important and endows an extra element of selectivity to the method.

5.8.2. Laboratory Procedure

The polynuclear aromatic hydrocarbons were obtained from specialist chemical suppliers in a high enough state of purity to require no further treatment. Spectroscopic grades of acetone and methanol were quite satisfactory and 1,2-dibromoethane was also used 'as received' because it decomposes on redistillation to produce a fluorescent product. Dibromoethane is suspected to be a potential carcinogen and so great care must be taken when handling it.

α- and β-cyclodextrins were recrystallised from boiling water but γ-cyclodextrin was used as received. The water used was deionised and triply distilled. All glassware was washed with spectroscopic grade acetone and methanol and dried at 100 °C in an oven before use.

Stock solutions of the analytes were made up as follows:

An aliquot portion of the stock solution of the analyte of interest was transferred to a 10 cm^3 volumetric flask and evaporated gently on a hotplate to remove the solvent. 1,2-dibromoethane (0.1 to 0.5 cm^3) was then added followed by the cyclodextrin solution to bring the volume up to the mark. The solution was then shaken vigorously by hand to form the complex. The solutions were generally slightly cloudy due to the formation of a precipitate containing excess of the dibromoethane but this did not adversely affect phosphorescence measurements. The prepared sample was transferred to a standard fluorescence cuvette equipped with a Teflon stopper de-aerated by bubbling ultra high purity nitrogen through it for 15 minutes. The cuvette was then tightly stoppered and transferred to the spectrometer for phosphorescence measurement.

Π The instrument used in this work was a Spex Fluorolog 2+2 dual double monochromator spectrofluorimeter (ie two monochromators in series are used both on the emission side and the excitation side of the sample (hence the '2+2' designation). This arrangement ensures high spectral purity of the radiation and very low stray light levels. The instrument has a 450 watt xenon continuous light source, continuously variable slits giving a bandwidth of 1.8 nm per mm and a cooled R928 photomultiplier tube detector which minimises the noise level. Altogether this instrument has a far higher performance than any we have previously encountered. What feature of this instrument makes it possible to observe the fluorescence and phosphorescence from turbid samples without difficulty?

The very low stray light specification made possible by the '2+2' monochromator and the high dispersion. This is in fact a 'total luminescence' spectrometer since the source is continuous and the fluorescence, phosphorescence and scattering enter the first emission monochromator simultaneously. An instrument equipped with time resolution using a pulsed or interrupted source would not in fact require such a high optical specification and would therefore be much cheaper. It would give equivalent performance when measuring the phosphorescence but fluorescence measurement would be seriously affected by the stray light generated by the Tyndall scattering of the precipitate.

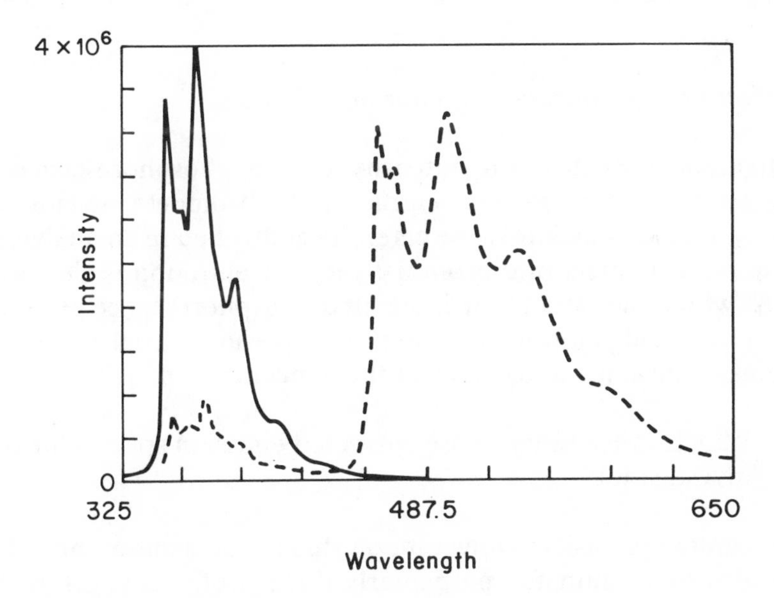

Fig. 5.8a. *Emission spectra of phenanthrene (5 × 10^{-5} mol dm^{-3})*

—— - in β-cyclodextrin (10^{-2} mol dm^{-3})
- - - - in β-cyclodextrin (10^{-2} mol dm^{-3}) and dibromoethane (0.58 mol dm^{-3})

5.8.3. Results and Optimisation of the Procedure

The influence of the dibromoethane is clearly shown in the emission spectra of phenanthrene (5×10^{-5} mol dm^{-3}) in β-cyclodextrin (10^{-2} mol dm^{-3}) in Fig. 5.8a. In the absence of dibromoethane the emission is entirely fluorescence, showing the usual well-defined fine structure associated with this molecule, and maximum emission at 368 nm. In the presence of dibromoethane (0.58 mol dm^{-3}) the emission spectrum is dominated by the much broader, still structured, phosphorescence band with maximum emission at 500 nm.

The fluorescence is quenched by about 90% by the dibromoethane. This behaviour is entirely reproducible and the emission spectrum does not change with time. There was no evidence of photochemical degradation of dibromoethane or the cyclodextrin complex on prolonged exposure.

(*a*) *Effect of Concentration of Dibromoethane*

As might be expected, the intensity of the phosphorescence is strongly dependent on the concentration of dibromoethane and this is clearly a factor which must be carefully controlled in an analytical procedure. The effect of dibromoethane concentration is shown in Fig 5.8b where the ratio of the integrated phosphorescence intensity to the integrated fluorescence intensity, measured from corrected emission spectra, is plotted against the concentration.

Π Why is it necessary to use corrected emission spectra for this investigation?

The integrated intensity involves measuring the area under the band. If this is to be meaningful, particularly if the profile changes as the concentration changes, account must be taken of the variation of detector response with wavelength because the phosphorescence band is very broad and covers a range of wavelengths where the photomultiplier response is falling off very rapidly.

Π What is the most notable feature of the graph shown in Fig. 5.8b?

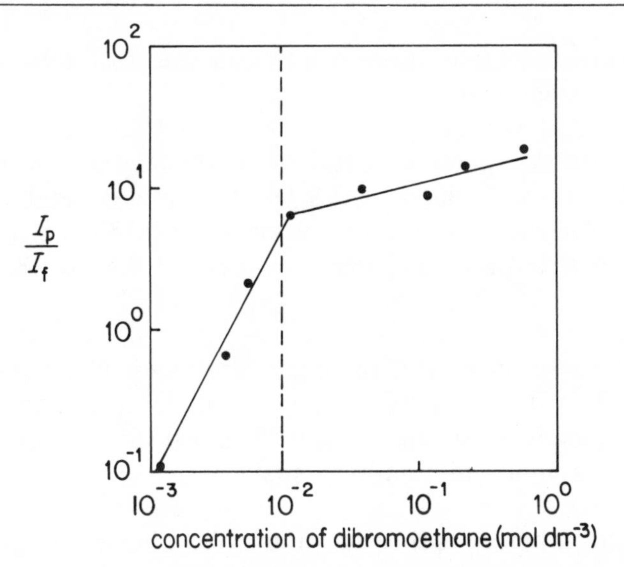

Fig. 5.8b. *The phosphorescence to fluorescence emission ratio of phenanthrene as a function of dibromoethane concentration*

The graph is essentially linear at high and low concentrations but there is sharp change in the slope in between.

Π At what concentration of dibromoethane does the change in slope occur? What is the significance of this concentration?

10^{-2} mol dm^{-3}.

This is the same as the molar concentration of β-cyclodextrin and we have the exact stoichiometric quantities of the two reagents to form a 1 : 1 complex.

Π On the basis of these results what quantity of dibromoethane would you recommend adding in an analytical procedure based on the phosphorescence measurement? What volume does this correspond to in the procedure described in Section 5.8.2? (The density of dibromoethane is 2.18 g cm^{-3}).

Enough to make the concentration of dibromoethane in the sample 0.1 mol dm^{-3}; a 10-fold molar excess. This is well up the higher section of the graph where the slope is much less. 0.086 cm^3 is required to give an equimolar quantity to the β-cyclodextrin in 10 cm^3

of the 0.01 mol dm^{-3} solution, which contains 10/1000 × 0.01 = 10^{-4} moles of cyclodextrin.

1 mole of 1,2-dibromoethane, $CH_2Br.CH_2Br$, weighs (2 × 12.01) + (4 × 1.008) + (2 × 79.90) = 187.8 g so 10^{-4} moles weigh 0.0188 g. A 10-fold molar excess would therefore weigh 0.188 g. Since 1 cm^3 weighs 2.18 g, 0.188 g would have a volume of 0.188/2.18 = 0.086 cm^3.

This is at the bottom end of the range specified in the procedure (0.1 – 0.5 cm^3). In fact, the authors maintained the final dibromoethane concentrations between 0.23 and 0.58 mol dm^{-3} in all the samples for which they publish spectra.

Π By how much does the ratio of the phosphorescence intensity to fluorescence intensity rise if the dibromoethane concentration is raised from 0.23 to 0.58 mol dm^{-3}?

1.26. In the upper section of the graph an increase of a factor of 10 in the concentration increases the log of I_p/I_f by 0.248. (Note that Fig. 5.8b is a log–log plot so the gradient relates to ratios rather than differences). Hence the range of concentration covers a factor of 0.58/0.23 = 2.52 which will increase the log of phosphorescence/fluorescence ratio by a factor of

$$[(\log 2.52/\log 10) \times 0.248 = 0.0994 \text{ or by } 1.26].$$

The increase in the ratio is caused by the fluorescence going down as well as by the phosphorescence going up. In practice, the analysis would be based on a direct measurement of phosphorescence since this is simpler and there is nothing to be gained by using the ratio. Consequently the effect of dibromoethane concentration would be even less.

(b) Effect of Concentration of β-Cyclodextrin

Now we must check how the concentration of β-cyclodextrin affects the phosphorescence intensity. This is shown in Fig. 5.8c in the more direct way since the ordinate is now the phosphorescence intensity. This curve has a rather more complex 'S-shape' but it too levels out appreciably at the high concentration end. The solubility of β-cyclodextrin prevents the use of concentrations much above 0.01 mol dm^{-3} so the authors settled on a value of 0.01 mol dm^{-3} as the working concentration.

Π By how much does the phosphorescence intensity increase if the concentration of β-cyclodextrin is raised by 25% from 0.008 to 0.01 mol dm^{-3}?

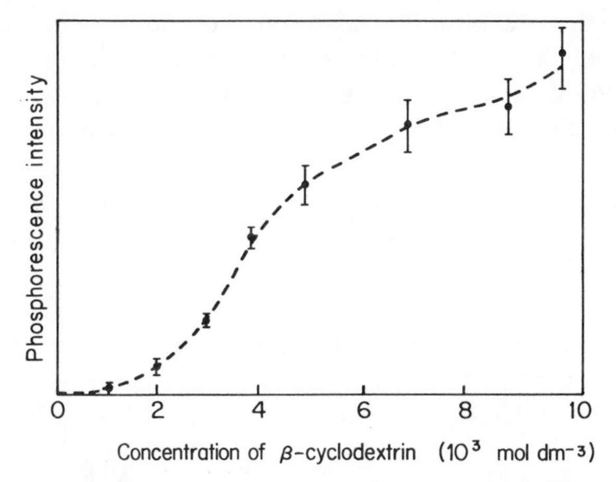

Fig. 5.8c. *Phosphorescence intensity of phenanthrene as a function of β-cyclodextrin concentration*

By 17% (a factor of 1.17). In practice there should be no problem in keeping the volume of the β-cyclodextrin added constant to much better than $\pm 10\%$ so the error in the phosphorescence reading from this cause will be negligible.

(*c*) *Effect of Specific Cyclodextrin Used*

There is yet another variable which we can manipulate to control the selectivity of the method. α, β and γ-cyclodextrins differ in the size of the lyophilic cavity into which the analyte is complexed. Clearly this has got to be large enough to accommodate the analyte molecule leaving enough room for the dibromoethane molecule in order to form the ternary complex. Fluorene for example is too large to allow the dibromoethane molecule to accompany it into the cavity of β-cyclodextrin (inner diameter 0.78 nm) and so no phosphorescence is observed (though the fluorescence is enhanced). With γ-cyclodextrin, however, there is intense phosphorescence as shown in Fig. 5.8d.

Similarly, phenanthrene, which phosphoresces strongly in β-cyclodextrin, gives only enhanced fluorescence in α-cyclodextrin (cavity diameter 0.6 nm) because there is no room for the dibromoethane molecule to be included as well. This size factor raises interesting possibilities for developing selective methods for particular compounds which are not available with micelles. Both naphthalene and phenylnaphthalene for example exhibit strong RTP in thallium dodecanoate but only naphthalene gives a phosphorescence signal in β-cyclodextrin.

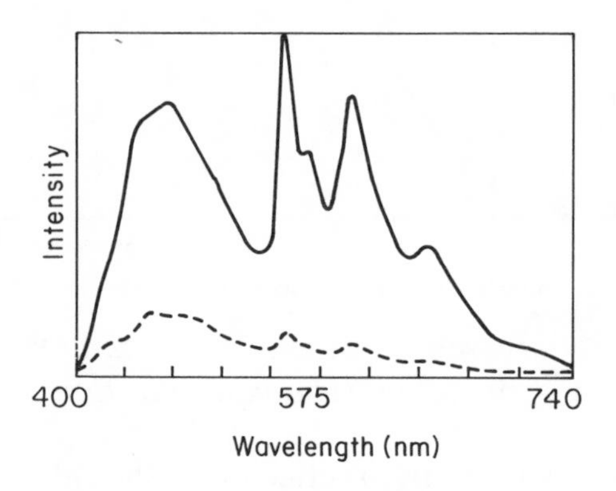

Fig. 5.8d. *Emission Spectra of Fluorene in*
——- γ-cyclodextrin (10^{-2} mol dm^{-3})
- - - β-cyclodextrin (10^{-2} mol dm^{-3})

(*d*) *Effect of Quenching Agents*

The fact that phosphorescence is observed at room temperature from molecules protected by micelles or the cyclodextrin structure is largely due to the prevention of mechanisms which might deactivate the triplet excited state. It is not surprising, therefore, to find that ionic species like the halide ions do not affect the emission. On the other hand it is still necessary with these liquid RTP procedures to de-oxygenate the sample and this step is indeed included in the present method. However the influence of oxygen is less marked with cyclodextrin and, although omission of the de-aeration step leads to a loss of 60–90% of the phosphorescence emission this is not serious in many cases because of the high intensity available. Where this step can be conveniently omitted, there will be a gain in speed of analysis and in precision. The effect of oxygen depends very much on the size of the analyte molecule. Thus, while the presence of oxygen reduced the emission of naphthalene to 8.9% of the value for the de-aerated sample, with 1-methyl-naphthalene 28.9% of the emission remains and other substituted naphthalenes also retain >20% of the emission.

5.8.4. Analytical Potential of the Cyclodextrin Procedure

Although no detailed results are given, the authors claim that high phosphorescence intensity is achieved for many polynuclear aromatic hydrocarbons in the nmol dm^{-3} range (18 compounds are listed with their excitation and emission wavelengths). In the case of phenanthrene and acenaphthene,

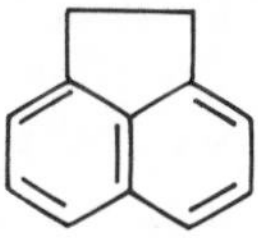

calibration graphs are linear over 4 orders of magnitude. At the bottom end of the range sensitivity is blank limited by fluorescent impurities in the cyclodextrin which cannot be removed by recrystallisation. Even so, the detection limits for phenanthrene and acenaphthene (based on a signal/noise ratio of 3 in this instance) are estimated to be 5×10^{-13} and 1×10^{-11} mol dm^{-3} respectively. The precision of measurements is typically 8–10%.

The phosphorescent lifetime of cyclodextrin solutions is in the microsecond region which is much longer than that of typical micelles. (Deoxygenated cyclodextrin samples often show visible emission for several seconds after removal from the excitation beam.)

This makes much lower demands on the instrumentation in time resolved systems.

Sample preparation time is comparable with other RTP methods but, if the de-aeration step is omitted the cyclodextrin procedure is significantly faster. This is not an option with the other technique and, in the case of micelles the de-aeration step causes additional problems by the formation of abundant soap suds. The possibility of omitting the de-aeration step makes it practicable to consider the application of the cyclodextrin technique to flow systems including LC detection. Here the use of time resolution to avoid background emission would give RTP a distinct advantage over other methods. The selectivity aspect of the cyclodextrin procedure could be advantageous but, of course, it does mean that the range of compounds to which it can be applied is limited compared with the other liquid phase RTP methods – you can't have it both ways! Unfortunately, the molecules of the higher members of the polynuclear aromatic hydrocarbon series such as benzpyrene (which are of considerable analytical importance because of their carcinogenic properties) and many important drugs are too large to fit into the cyclodextrin cavity. However, there are many compounds which can be induced to phosphoresce in cyclodextrin so the technique is likely to be a useful addition to the analyst's armoury. A subsequent paper by the same authors [*Anal. Chem.*, 1984, **56**, 331] lists 16 nitrogen heterocycles for which RTP has been observed in cyclodextrin. These compounds are also important in the drug industry and not easily accessible by conventional fluorescence methods because they often have very low fluorescence efficiencies.

SAQ 5.8a

RTP in both micelles and cyclodextrins requires the presence of a 'heavy atom'. Describe *three* ways in which this may be introduced, indicating in each case whether the method applies to micelles, cyclodextrins or both.

SAQ 5.8a

SAQ 5.8b

Which of the following items constitute an advantage of cyclodextrins over micelles in RTP of fluid samples?

(*i*) Very wide application.
(*ii*) No need for de-aeration before measurement.
(*iii*) Longer lifetime of phosphorescence.
(*iv*) Requires the presence of a heavy atom.
(*v*) More selective.

SAQ 5.8c

Phosphorescence can be observed in the presence of scattered radiation either by the use of a total luminescence spectrometer of high performance or a fluorescence spectrometer with a pulsed source. Explain the principles of these two methods.

SUMMARY AND OBJECTIVES

Summary

Fluorescence is used as an alternative, more sensitive technique to colorimetry in all forms of liquid chromatography. In tlc and paper chromatography spots are visualised by forming derivatives if components are not naturally fluorescent or by quenching of a fluorescent indicator incorporated into the stationary phase in tlc plates.

Limited possibilities of qualitative analysis are available with a special accessory to record the spectrum from a spot. Column chromatography can also make use of natural fluorescence or the formation of derivatives before or after separation of the components on the column. Reagents giving derivatives with similar fluorescence characteristics are most convenient, the selectivity being achieved by the chromatography.

Use is made of the high selectivity of the antibody/antigen reaction of immune systems by attaching a fluorescent label to the analyte (antigen). This competes with the analyte present in a sample of unknown concentration for a limited amount of antibody and a determination of free or bound labelled antigen (or the ratio) leads to the unknown concentration by fluoroimmunoassay. This technique is very similar to radioimmunoassay and, though less sensitive in some instances, it is also less expensive and avoids the problem of safe working associated with radioisotopes. The materials required for FIA require special production techniques but are normally provided for the analyst in a 'kit'.

To avoid the need to freeze samples in liquid nitrogen the technique of room temperature phosphorescence has recently been developed because of the availability of relatively low-cost instruments having pulsed sources and limited time resolution capabilities. Deactivation of the excited triplet state molecules responsible for the fluorescence is prevented either by examining the sample in the solid state or by absorption of the sample in the liquid phase into micelles such as thallium dodecanoate or large organic molecules such as cyclodextrins which have a similar structure with a hydrophobic region inside a hydrophilic shell. In both cases interference from scattering or fluorescence of the analyte (or other species) is avoided by the time resolution principle. With liquid phase room temperature phosphorescence (RTP) the presence of heavy atoms closely associated with the analyte is necessary to enhance the phosphorescence.

Four case studies are provided to illustrate the practical aspects of different aspects of luminescence analysis.

(*i*) Analysis of a mixture of four alkaloids utilising fluorescence of the analyte modified by chemical conditions and assisted by physical separation.

(*ii*) Analysis of selenium at trace levels in river water based on the formation of a fluorescent derivative with 2,3-diaminonaphthalene and utilising two different preconcentration procedures to allow a separate determination of Se(IV) and Se(VI).

(*iii*) A fluoroimmunoassay of progesterone in human serum using fluoresceinamine as a label in a heterogeneous assay involving sub-microgram quantities of the analyte.

(*iv*) Analysis of polynuclear aromatic hydrocarbons by RTP in cyclodextrin using 1,2-dibromoethane to enhance the phosphorescence which is measured in a high performance, total luminescence spectrometer.

Objectives

You should now be able to:

- describe the use of fluorescence for visualising spots in paper and thin layer chromatography;

- describe the method for obtaining luminescence spectra from spots on thin layer plates;

- describe the use of fluorescence in detecting components eluted from a liquid chromatograph;

- distinguish between pre-column and post-column derivatisation in liquid chromatography and compare their relative merits;

- quote examples of derivatising reagents;

- explain the principles of immunoassay and describe applications using fluorescent labels;

- compare the relative merits of RIA and FIA;
- distinguish between homogeneous and heterogeneous immunoassays;
- describe the technique of Room Temperature Phosphorescence as applied to solid samples;
- explain how phosphorescence can be observed at room temperature in liquid solution by the use of viscous media, micelles, and cyclodextrins;
- extract the salient information from a published analytical method and translate this into a practical laboratory procedure;
- describe standard laboratory procedures used in luminescence methods and point out the precautions necessary to achieve successful results with both organic and inorganic materials;
- manipulate data produced in a photoluminescence procedure and calculate and interpret the result of a specific analysis from raw data.

Self Assessment Questions and Responses

SAQ 1.1a

Select a source of light from the list below which is an example of

(*A*) radioluminescence

(*B*) photoluminescence

(*C*) bioluminescence

(*D*) chemiluminescence

Sources:

(*i*) cold light sticks.

(*ii*) a theatrical mask which glows on a darkened stage.

(*iii*) glow-worms.

(*iv*) luminous paint.

Response

(*A*) – (*iv*)
(*B*) – (*ii*)
(*C*) – (*iii*)
(*D*) – (*i*)

****

SAQ 1.2a

(*i*) The most probable electronic transition in the benzene molecule is the 0,1 transition. What is the wavelength of the absorption corresponding to this transition? Refer to Fig. 1.2b.

(*ii*) What are the wavelengths of the 0,0 and the 0,3 bands?

(*iii*) Which of these bands corresponds to the smallest transition energy?

(N.B. The weak band at 268.2 nm involves one of the other vibrations of benzene which we have ignored in our discussion.)

There is some ambiguity in the use of the word 'band' in discussing uv spectra. The term '0,1 band' refers to one specific feature in the fine structure of the overall 'electronic band'. The meaning is generally clear from the context in which it is used.

Response

(*i*) 254.6 nm. The strongest band corresponds to the most probable transition.

(*ii*) 0,0 at 260.8 nm; 0,3 at 243.2 nm.

(*iii*) 0,0 is of lowest energy.

****

SAQ 1.2b

Choose words from the following list to fill the blanks in the paragraph below:

longer shorter higher lower

(each word may be used once, twice, or not at all)

'The most important feature of the transitions giving rise to the fluorescence band is that they are of ________ energy than those associated with the absorption band and so the arrows are ________ on the energy scale of Fig. 1.2a. Consequently, the fluorescence emission band appears at ________ frequency and therefore at ________ wavelength than the absorption band'.

Response

lower shorter lower longer (in that order).

****

SAQ 1.2c

Draw a simple energy level diagram to show the ground and first excited electronic states of a molecule with *five* vibrational levels in each state.

(*i*) Label the vibrational levels with the quantum number v′ or v″.

(*ii*) Identify the level in which most molecules will be found at room temperature.

(*iii*) Identify the level from which fluorescence emission is most likely to originate.

(*iv*) Draw an arrow to show the transition giving rise to the 0,4 absorption band.

(*v*) Draw an arrow to show the transition giving rise to the 0,1 emission band.

(*vi*) Which band, (*iv*) or (*v*), has the longer wavelength?

Response

The diagram is based on Fig. 1.2a (Note that although the vibrational levels are shown as equally spaced, they actually converge slowly as v increases.)

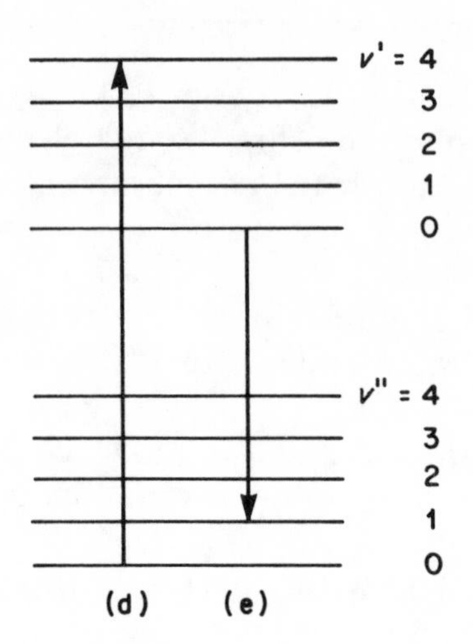

(*i*) see diagram

(*ii*) v″ = 0

(*iii*) v′ = 0

(*iv*) and (*v*) see diagram – make sure that you drew the arrows the right way round!

(*vi*) The 0,1 band has the longer wavelength – the arrow is shorter, the transition energy less, the band frequency lower and so the wavelength longer.

Note that in describing the emission transition as '0,1' we place v′ first. In specifying absorption transitions we place v″ first. This system is in fact consistent provided that you remember that the quantum number of the *initial* state is placed first.

****

SAQ 1.2d

Which transition gives rise to a band in both the excitation and emission spectrum of a fluorescent compound?

Response

The 0,0 transition. All other transitions which excite the molecule are of greater energy so the excitation spectrum extends from the 0,0 band to shorter wavelength. All other transitions associated with the emission of radiation are of lower energy so the emission spectrum extends from the 0,0 band to longer wavelength.

****

SAQ 1.2e

Evaluate the Stokes' Shift for anthracene (refer to Fig. 1.2c).

Response

45 nm. This is the difference between the maxima of the emission band, 400 nm, and the excitation band, 355 nm.

****

SAQ 1.3a

Which of the following times would you consider to be reasonable to allow adjustment of the solvent molecules in the sequence of Fig. 1.3a to take place:

10^{-20} s, 10^{-15} s, 10^{-10} s, 10^{-5} s ?

Response

10^{-10} s.

This time is intermediate between the time for absorption (10^{-15} s) and the lifetime of the excited state (10^{-8} s). 10^{-5} s is obviously too long since the photon would have been emitted before the adjustment of solvent molecules had taken place. 10^{-15} s is much too short a time for for any significant change in molecular position to occur. 10^{-20} s corresponds to a minute fraction of the time period of oscillation of the electromagnetic field of uv radiation and is too short even for the transition to occur.

****

SAQ 1.3b

In the diagram below, the central shaded ellipse represents an excited molecule with its polarity indicated by + − at the centres of electrical charge. The surrounding unshaded ellipses represent polar solvent molecules.

⟶

SAQ 1.3b (cont.)

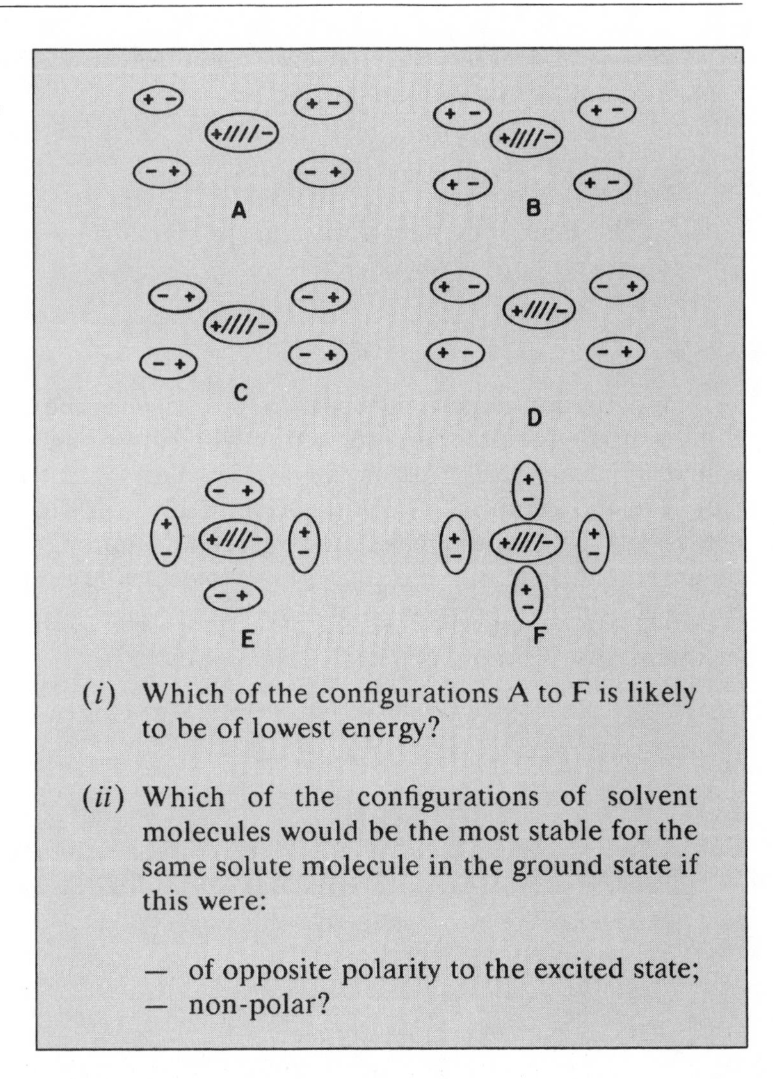

(*i*) Which of the configurations A to F is likely to be of lowest energy?

(*ii*) Which of the configurations of solvent molecules would be the most stable for the same solute molecule in the ground state if this were:

— of opposite polarity to the excited state;
— non-polar?

Response

(*i*) B. In this configuration the solvent molecules are lined up with their negative ends closest to the positive end of the solute molecule and *vice versa*. In other configurations some solvent molecules present a 'like' charge to the solute (+ to +). E is a possible exception but is probably not as favourable as B because there is effectively no interaction with the end-on solvent molecules, the 'like-like' repulsion cancelling out the 'like-unlike' attraction.

(*ii*) — of opposite polarity to the excited state:

C. The same argument applies as in (*i*) but now of course all the charges are reversed.

— non polar:

No configuration is particularly favoured because there can be no dipole–dipole (coulombic) interaction with solvent molecules if the solute molecule is non-polar. In fact, solvent–solvent interactions would now predominate so probably B, C or F, in which these are most favourable, would make the largest contribution to the average environment. In the liquid phase the molecules are relatively free to move from one position to another.

****

SAQ 1.4a

> The values of ϕ_f for three organic compounds A, B and C are 0.55, 0.91 and 0.12 respectively. Using this information only, which is the most strongly fluorescent?

Response

B. A high value of ϕ_f close to 1.0 indicates that a high proportion of absorbed photons are re-emitted.

(The values quoted are in fact those for quinine sulphate, and the sodium salts of fluorescein and tryptophan respectively. The fluorescence intensity also depends on the absorptivity as we shall discover in Part 3.)

****

SAQ 1.4b

State whether each of the following statements is true or false:

(*i*) Molecular fluorescence competes with two radiationless processes: Internal Conversion, Inter-system Crossing.

(*ii*) An excited molecule with a fused ring structure will readily lose its energy by internal conversion.

(*iii*) In order for inter-system crossing to be effective it must occur within 10^{-8} s of the molecule being excited.

(*iv*) A molecule in which inter-system crossing is very efficient is unlikely to be fluorescent.

(*v*) A molecule can pass from a $\pi \rightarrow \pi^*$ excited state to an $n \rightarrow \pi^*$ excited state by internal conversion.

(*vi*) Molecules in $\pi \rightarrow \pi^*$ excited states are more likely to undergo inter-system crossing than those in $n \rightarrow \pi^*$ states.

⟶

SAQ 1.4b (cont.)

(*vii*) The fluorescence efficiency of a compound varies when it is dissolved in different solvents as a result of differences in the effectiveness of quenching.

(*viii*) Energy is removed very efficiently from the excited state of compounds containing bromine by inter-system crossing.

(*ix*) De-excitation following inter-system crossing is achieved primarily by vibrational relaxation.

Response

(*i*) True. The effectiveness of the two radiationless processes in competition with fluorescence emission governs the extent to which a compound fluoresces.

(*ii*) False. The fused ring structure will make the molecule more rigid which will inhibit transfer of energy through the vibrational modes. Molecules of this type are usually fluorescent.

(*iii*) True. The lifetime of the $v' = 0$ level of the initial excited state is 10^{-8} s so unless inter-system takes place within this time a photon will be emitted. The reciprocal of the time taken for the molecule to cross over to the lower state is the *rate constant* for the process. We may therefore say that, if a molecule is to fluoresce the rate constant for fluorescence must be greater than those for inter-system crossing and internal conversion. In a complex situation in which excited molecules may fluoresce, decompose or lose their energy in other ways it is often useful to discuss the processes in terms of rate constants – as one would for competing chemical reactions.

(*iv*) True. Efficient inter-system crossing rapidly depopulates the state from which fluorescence would occur.

(*v*) True. This is the normal procedure by which a molecule drops to its lowest excited state when that state is $n \rightarrow \pi^*$.

(*vi*) False. Inter-system crossing is much more likely from an $n \rightarrow \pi^*$ state because it has a much longer lifetime.

(*vii*) True. Quenching commonly occurs because of an external process and solvent molecules are an external influence. The solvent has a much less marked effect upon internal transfer of energy.

(*viii*) False. There is no change in the energy during inter-system crossing, deactivation occurs after the molecule arrives in the lower state. The presence of the bromine atom will, of course, make inter-system crossing more efficient.

(*ix*) False. Vibrational relaxation will only take the molecule as far as the $v' = 0$ level of the lower excited state. It gets to the ground state by a quenching mechanism. Only then will it be de-excited.

****

SAQ 1.4c

What three differences would you expect to find in the fluorescence spectrum of a compound showing vibrational fine structure if the solvent were changed from cyclohexane to ethanol?

Response

(*i*) The vibrational fine structure would become less well resolved in ethanol because it is more polar than cyclohexane.

(*ii*) The difference between the wavelengths of the 0,0 band in the excitation and emission spectra would be greater in ethanol.

(*iii*) The overall intensity for the same concentration would differ because of the different effects of the two solvents on ϕ_f.

****

SAQ 1.5a

The energy level diagram given below shows some of the lower vibrational levels of a ground electronic state, G, and two excited electronic states, S and T.

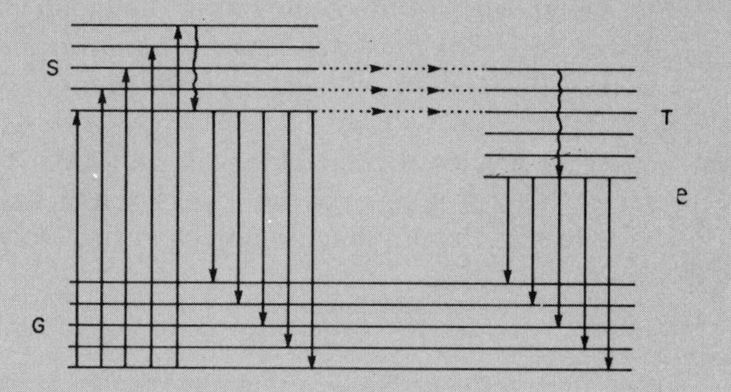

There are six different processes represented by arrows of various types. Label them A to F to correspond with the following descriptions:

A Excitation
B Fluorescence emission
C Phosphorescence emission
D Vibrational relaxation of the upper excited state
E Vibrational relaxation of the lower excited state
F Inter-system crossing

Response

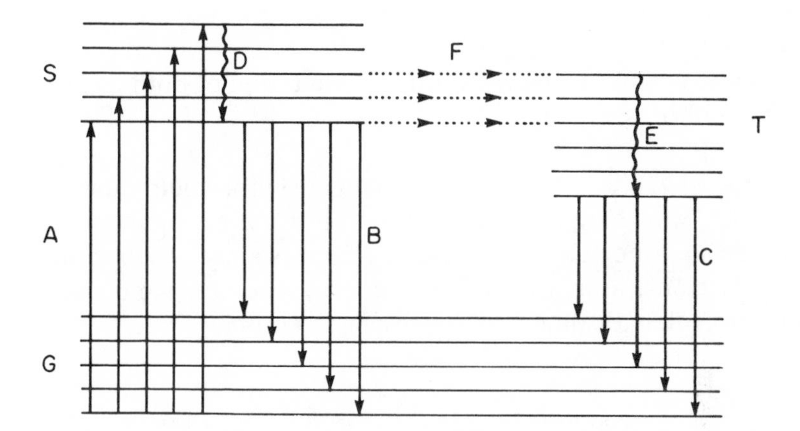

Straight, vertical arrows represent transitions associated with the absorption or emission of photons. Curly or dotted arrows represent radiationless transfer of energy. The processes represented by D and E occur within a single system of vibrational levels associated with one electronic state and are examples of 'vibrational relaxation'. These differ from inter-system crossing represented by F where the molecule transfers from one set of vibrational levels to another associated with a different electronic state.

****

SAQ 1.5b

> Place the following bands observed in the photoluminescence spectrum of a single compound in order of increasing wavelength:
>
> fluorescence emission
> fluorescence excitation
> phosphorescence emission

Response

excitation < fluorescence emission < phosphorescence emission

(The same process is involved for the excitation of both fluorescence and phosphorescence).

In symbols, $\lambda_{ex} < \lambda_{fl} < \lambda_{ph}$, a single symbol being used for the excitation wavelength. The arrows representing phosphorescence emission are shorter than those for fluorescence. Hence phosphorescence transitions are of lower energy and the band appears at longer wavelength than the fluorescence emission band.

****

SAQ 1.5c

The diagram below shows how the intensity of photoluminescence emission, excited by a single 'pulse' of uv radiation, varies with time. Of the two curves (full line and dotted) which corresponds to fluorescence and which to phosphorescence?

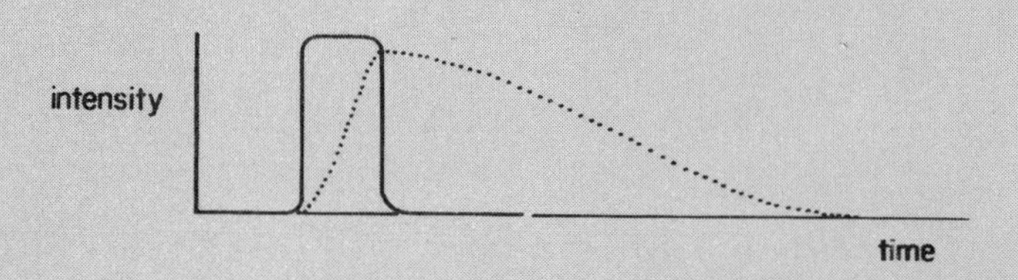

Indicate on the diagram the time at which

(*i*) the pulse was switched on, by t_1;
(*ii*) the pulse was switched off, by t_2;
(*iii*) the phosphorescence could be measured without interference from fluorescence, by t_3.

Response

The full line is fluorescence and the dotted line is phosphorescence because the latter is of longer duration.

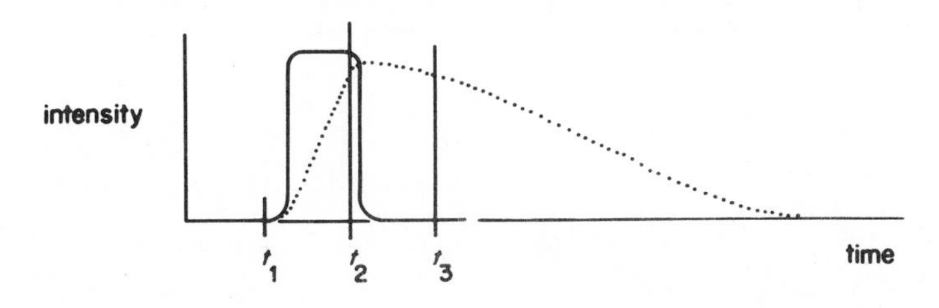

We assume here that the fluorescence follows the profile of the exciting pulse closely with a delay of about 10^{-8} s which is negligible on the time scale of our diagram. The relative intensities of fluorescence and phosphorescence vary widely with different molecules and are often reversed.

****

SAQ 1.5d

> What radiation processes compete with
>
> (*i*) fluorescence emission
> (*ii*) phosphorescence emission
>
> as the means of deactivating an excited molecule?

Response

(*i*) Internal conversion and inter-system crossing.

(*ii*) Quenching is by far the most important.

****

SAQ 1.5e

How could you check whether long-lived emission from a sample frozen at 77 K was phosphorescence or delayed fluorescence?

Response

Melt the sample and record the fluorescence emission spectra. If this has the same wavelength as the long-lived emission the latter is delayed fluorescence. If the wavelength of the long-lived emission is longer it is phosphorescence.

****

SAQ 1.6a

Which of the statements (*i*) to (*v*) correctly completes the following sentence?

Transitions from T_1 to S_0 are forbidden because

(*i*) the state is depopulated by inter-system crossing before the transition can occur.
(*ii*) T_1 is of lower energy than S_1.
(*iii*) they involve a change in the electron spin of the molecule.
(*iv*) the spin of the T_1 state is zero.
(*v*) the probability of singlet-triplet transitions is very low.

Response

(*iii*) is correct.

(*i*) is wrong because, although inter-system crossing can take place from T_1 to S_1, it involves an increase in the energy and so proceeds at a much slower pace than S_1 to T_1. Hence it cannot result in a depopulation of the triplet state.

(*ii*) is wrong because, although T_1 is of lower energy than S_1, this does not determine whether or not a transition is allowed.

(*iv*) is wrong because the spin of the triplet state molecule is not zero. (The net spin of two parallel electrons is $\frac{1}{2} + \frac{1}{2} = 1$).

(*v*) is simply another way of saying that the transition is forbidden; it is not the reason.

****

SAQ 2.2a

Explain why the high pressure xenon arc is more satisfactory as a source for fluorescence work than either the tungsten filament lamp, the deuterium lamp or the mercury discharge lamp. When might the mercury lamp be a convenient alternative?

Response

The xenon arc is a continuous source of high intensity offering the operator a free choice of excitation wavelength. The tungsten lamp does not emit radiation below 320 nm and has a very low intensity outside the visible region. The tungsten lamp is also a continuous source of uv radiation but its intensity is too low for fluorescence work. The mercury lamp is a line source with very high intensity at certain wavelengths. If the excitation band of the sample falls between two mercury lines, the mercury lamp will be much less efficient than the xenon arc. It is, of course, not possible to obtain the excitation spectrum of a sample with a mercury source since no significant emission will be observed at excitation wavelengths between the mercury lines (though high pressure lamps do produce some band emission in these regions).

The mercury lamp becomes a viable alternative to the xenon arc for compounds which can be excited efficiently at one of the available wavelengths. Since the output is concentrated in relatively few atomic lines, the mercury lamp can be operated at a much lower power and yet provide the same intensity as the xenon arc at the wavelengths of the mercury lines. It is also possible to produce essentially monochromatic radiation with the combination of a mercury lamp and an interference filter. This is less expensive than a monochromator and also more efficient in terms of the energy reaching the sample. There is also less risk from explosion with low pressure versions of the lamp.

****

SAQ 2.3a

A spectrometer has a stray light level of 0.01%. In a particular experiment the value of I_0 was 1000 arbitrary units. The sample absorbed 1% of the incident radiation and scattered a further 1%. Calculate the value of I_f which is equal to the stray light level, using 180° geometry.

Response

0.0980 arbitrary units.

If the sample absorbs 1% of the incident radiation and 1% is scattered, 98% is transmitted – ie 980 arbitrary units. 0.01% appears as stray light in the monochromator, ie 0.098 units. The fluorescence intensity equal to the stray light therefore is approximately 10^{-4} I_o.

****

SAQ 2.3b

> Calculate the value of I_f which is equal to the stray light level, using 90° geometry for the same instrument as in SAQ 2.3a. Assume that 1% of the total scattered light enters the monochromator.
>
> Hence estimate the gain in sensitivity of the 90° configuration over the 180° or this sample.

Response

0.0000100 arbitrary units.

The total scattered radiation is 1% of I_o, ie 10 units. Of this, 1% enters the monochromator, ie 0.100 units. 0.01% of this appears as stray light = 0.0000100 units = 10^{-8} I_o.

The gain in sensitivity is of the order of 10^4.

The intensity of fluorescence entering the monochromator will be the same for both configurations assuming that the fluorescence distribution is the same in all directions (which is reasonable for liquid samples). It will therefore be possible to observe the fluorescence at much lower intensities with 90° geometry before it is obscured by stray light. This is very important in practice and is one of the main reasons why fluorescence is one of the most sensitive of analytical techniques.

****

SAQ 2.4a

Suggest a situation in which it would *not* be possible to use

(*i*) a glass cell
(*ii*) a silica cell
(*iii*) a polystyrene cell

for making fluorescence measurements on a liquid sample.

Response

(*i*) Measurements where the excitation wavelength is below about 320 nm since glass absorbs strongly in this region. Silica must be used in this situation.

(*ii*) Measurements on solutions containing hydrofluoric acid which etches the surface of both silica and glass. Otherwise silica can be used in almost all situations since it transmits down to 185 nm. It is however the most expensive of these materials.

(*iii*) Measurements on solutions in organic solvents such as acetone or chloroform. Polystyrene cells are useful for aqueous solutions containing HF though, like glass, they do not transmit below 300 nm. If you had to measure a solution containing HF below 300 nm you would have to use a cell made of polyethylene or teflon though these materials have optical properties which are far from ideal for accurate work.

****

SAQ 2.4b

How would

(*i*) a scratch, and
(*ii*) a fingerprint

on the optical surfaces of a cell affect the measurement of fluorescence?

Response

(*i*) Scratches scatter the radiation and if the scratched face were on the incident side of the cell or directly opposite to it some of the scattered exciting radiation would enter the monochromator leading to an increase in stray light. A scratch on the other faces has less effect.

(*ii*) Fingerprints leave greasy deposits which both absorb uv radiation and fluoresce. This causes trouble at low wavelengths due to loss of the fluorescence signal and an increase in the cell blank.

****

SAQ 2.5a

If you wanted to buy a filter/monochromator instrument to record the emission spectrum of a sample, which of the monochromators would be replaced by the filter?

Response

The excitation monochromator. To obtain the emission spectrum we excite the sample at fixed wavelength and scan the emission monochromator. Using a filter on the excitation side of the sample would be a satisfactory alternative.

****

SAQ 2.5b

Explain how you would determine the wavelengths of maximum excitation and emission for a compound for which they were not already known. You may assume that the emission monochromator of your dual monochromator spectrofluorimeter has a 'direct light' setting in which the grating acts as a mirror to reflect all the emitted radiation on to the detector.

Response

Set the emission monochromator to the 'direct light' position. Scan the excitation monochromator through its entire range at a fast speed (manually if possible, though modern instruments don't allow for this!) and observe the wavelength at which the maximum reading occurs on the read-out device. (If the fluorescence is in the visible you can judge this quite accurately by eye.)

Set the excitation monochromator to the wavelength to give maximum emission intensity and scan the emission monochromator rapidly to the end of its range, starting at a wavelength close to that of the exciting radiation.

Read the emission maximum accurately from this spectrum, set the emission monochromator close to this and record the excitation spectrum. Read the wavelength of maximum excitation accurately from this spectrum.

****

SAQ 2.6a

> What instrumental factors affect the appearance of fluorescence spectra, and make correction necessary when comparing spectra from different instruments? Distinguish between those factors which affect the excitation spectrum and those which affect the emission spectrum.

Response

Variation of source output, detector sensitivity and monochromator efficiency with wavelength. The source and excitation monochromator factors affect the excitation spectrum,and the detector and emission monochromator factors affect the emission spectrum.

****

SAQ 2.6b

> When is it necessary to correct fluorescence spectra? For what purposes is correction unnecessary?

Response

Excitation spectra must be corrected if they are to be compared with absorption spectra. Both excitation and emission spectra must be corrected if spectra from different fluorescence spectrometers are to be compared.

Correction is needed when fundamental fluorescence properties of a molecule such as fluorescence efficiency are to be determined.

Correction is unnecessary for routine fluorimetric analysis at fixed wavelengths.

****

SAQ 2.6c

What is a quantum counter? Explain how it is used to correct excitation spectra.

Response

A solution of a compound such as Rhodamine B in glycerol which emits radiation of intensity proportional to that of the incident radiation and independent of its wavelength. A small proportion of the incident radiation is reflected with a beam splitter on to the quantum counter. This emits radiation on to a photomultiplier which produces a reference signal proportional to the intensity of the incident radiation. The ratio of the fluorescence signal from the sample to this reference signal is computed and plotted against the excitation wavelength to give the corrected spectrum.

****

SAQ 2.7a

How do the characteristics of the radiation entering the monochromator in a phosphorescence instrument equipped with a pulsed source and electronic timing mechanism differ from that in an instrument with a continuous source and a rotating can or shutter system?

Response

With electronic gating all the scattered and phosphorescent radiation enters the monochromator. With the mechanical systems the phosphorescence is observed separately while the exciting radiation is cut off.

****

SAQ 2.7b

In what respects does the performance of the detector and the monochromator used for phosphorescence work need to be superior to the minimum required for fluorescence work?

Response

The detector and its associated electronic circuitry must have a very fast response/recovery time. The monochromator must have a very low stray light specification.

Although the monochromator is set to receive the phosphorescence at a different wavelength from that of the excitation radiation, some of the fluorescence and scattered radiation will emerge through the exit slit as stray light. If this is of high intensity compared with phosphorescence (as it would be with a solid sample) it will produce a large initial peak in the output. The detector would take a significant time (relative to phosphorescent life-times) to recover from this before the phosphorescence signal could be extracted – effectively imposing a lower limit on t_d. With modern UHF radio techniques it is not difficult to achieve response times of less than 1 μs.

****

SAQ 3.2a

Calculate the limiting concentration for linearity in mg dm^{-3} for the following compounds (using 1 cm cells):

		M_r	$\epsilon/dm^3\ mol^{-1}\ cm^{-1}$
(*i*)	β-carotene	534	122 000
(*ii*)	Naphthalene	128	5 600
(*iii*)	Benzene	78	200
(*iv*)	Chrysene	228	139 000

Response

(*i*) 0.088 mg dm^{-3}
(*ii*) 0.46 mg dm^{-3}
(*iii*) 7.8 mg dm^{-3}
(*iv*) 0.033 mg dm^{-3}

These are the concentrations which give an absorbance value of 0.02. Using the Beer–Lambert law, for β-carotene

$$A = \epsilon c d$$

$$= 122\,000 \times c \times 1$$

$$= 0.02$$

$$c = 0.02/122\,000$$

$$= 1.64 \times 10^{-7} \text{ mol dm}^{-3}$$

$$= 8.75 \times 10^{-5} \text{ g dm}^{-3}$$

$$= 0.0875 \text{ mg dm}^{-3}$$

Similarly for the other compounds.

Note that the limit is lower with compounds of low molecular mass and high absorptivity. (This also applies to the detection limit which is about 1000 times lower in each case.) In practice these effects tend to compensate to some extent since the absorptivity often increases with molar mass.

****

SAQ 3.2b

Identify the phrase which correctly completes the following sentence:

The chief objection to the use of the term 'parts per million' for solutions of low concentration is that

(*i*) it gives unreasonably high numerical values.
(*ii*) it is only valid for solutions of density 1.000 g cm^{-3}.
(*iii*) the million has a different meaning in different countries.
(*iv*) it is not possible to relate it to concentrations in w/v units even when the density is 1.000 g cm^{-3}.

Response

(*ii*) is correct.

The other statements are not true because, in (*i*) and (*iv*), the numerical values are the same as for the mg dm^{-3} unit and, in (*iii*), the million always means 1 000 000, unlike the billion which has two possible interpretations.

****

SAQ 3.2c

What additional objections are there to the unit 'parts per trillion' and its abbreviation 'ppt' besides its dependence on density?

Response

The term 'trillion' is ambiguous meaning 10^{12} or 10^{18}. The abbreviation is also used for 'parts per thousand'.

****

SAQ 3.2d

In a particular fluorimetric analysis the limit of detection was 0.35 ng cm^{-3} and the upper limit of concentration was 50 mg dm^{-3}. What is the dynamic range of this analysis?

(*i*) 140 000
(*ii*) 14 000
(*iii*) 1400
(*iv*) 14 285
(*v*) 0.00007

Response

(*ii*) is correct.

The dynamic range is 50 mg dm^{-3}/3.5 ng cm^{-3} ie highest concentration/lowest concentration.

The lowest measurable concentration is taken as 10 times the detection limit, $10 \times 0.35 = 3.5$ ng cm^{-3}. We need to express both concentrations in the same units – μg cm^{-3} is the obvious choice. This gives

$$\text{Dynamic range} = 50\ \mu\text{g cm}^{-3}/0.0035\ \mu\text{g cm}^{-3}$$

$$= 14\,000$$

If you got (*iv*), you did the calculation correctly but omitted to round off the answer shown on your calculator to the correct number of digits. Both the detection limit and the highest usable concentration are very approximate estimates.

If you got (*i*), you probably used the detection limit as the lowest measurable concentration. In practice it is generally inadvisable to use any analytical technique down to its detection limit.

If you got (*iii*), you made a mistake with the power of 10 – a very common error in handling low concentrations! Any other answer starting with '14' would also indicate an error in the power of 10. If you got (*v*), you had the fraction the wrong way up – the magnitude of the result should have warned you that something was amiss! Any other value with 7 as the first significant digit means that there is an error in the power of 10 as well.

****

SAQ 3.2e

Are the following statements true or false?

(*i*) The concentration should always be plotted as the abscissa of a calibration graph.

(*ii*) The expression for the fluorescence intensity is $I_f = \phi_f I_0(1 - e^{kcd})$

(*iii*) The fluorescence intensity is proportional to the concentration provided ϵcd does not exceed 0.05.

(*iv*) The limit of detection for fluorescence is usually about 10 times larger than that for absorption spectroscopy.

(*v*) In the linear region of a fluorescence calibration graph the slope is equal to $\phi_f \epsilon$ for a particular instrument.

Response

(*i*) True. The analytical signal depends on the concentration and so it should be plotted as ordinate.

(*ii*) False. The exponential index has a negative sign $-e^{-kcd}$.

(*iii*) False. ϵcd is the absorbance which should not exceed 0.02.

(0.05 is the limit for 2.303 ϵcd.)

(*iv*) False. It is the limit of linearity of the fluorescence graph which is referred to. The detection limit is some 1000 times lower.

(*v*) False. The slope is $\phi_f\ I_0\epsilon$. We shall see later that other factors have to be included when comparing results from different instruments – or even the same instrument under different conditions.

****

SAQ 3.3a

What would be the effect on the fluorescence reading of diluting, by a factor of 2, the following three quinine solutions?

(*i*) 0.5 mg dm^{-3}
(*ii*) 200 mg dm^{-3}
(*iii*) 80 mg dm^{-3}

Refer to the two calibration curves for quinine sulphate in Fig. 3.3f.

⟶

SAQ 3.3a (cont.)

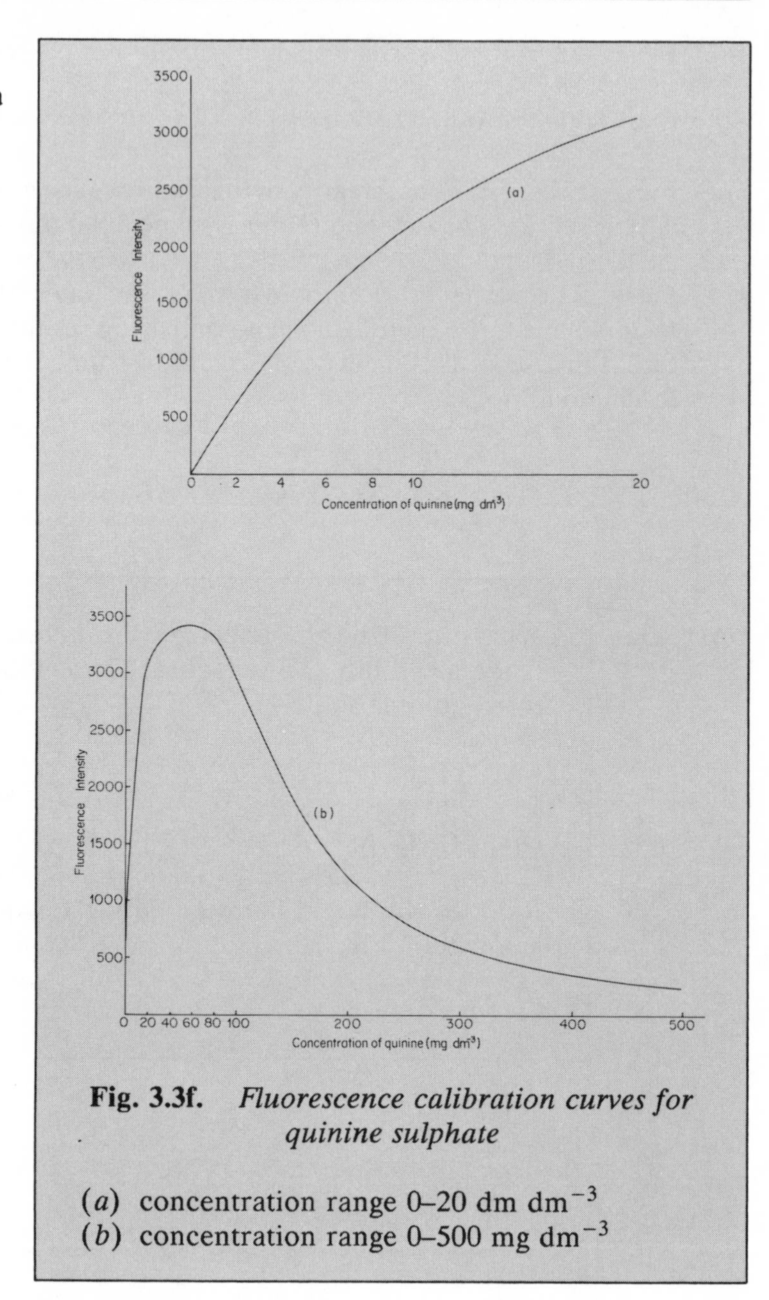

Fig. 3.3f. *Fluorescence calibration curves for quinine sulphate*

(a) concentration range 0–20 dm dm^{-3}
(b) concentration range 0–500 mg dm^{-3}

Response

(*i*) The reading would decrease by a factor of about 2 since this concentration is in the 'normal calibration range'.

(*ii*) The reading would increase substantially from 1200 to 3000, since the concentration of 200 mg dm^{-3} is well above that for maximum fluorescence intensity.

(*iii*) The reading would not alter very much since the concentration is just above the maximum, and the concentration of 40 mg dm^{-3} just below that for maximum fluorescence intensity.

****

SAQ 3.3b

Are the following statements true or false?

(*i*) There are two contributing factors that give rise to curvature of a fluorescence intensity versus concentration calibration graph. They are the inner filter effect and the dependence of fluorescence intensity on e^{-kcd}.

(*ii*) Using 90° geometry, the maximum observed in the fluorescence intensity versus concentration calibration graph is due to the dependence of fluorescence intensity on e^{-kcd}.

(*iii*) With 180° geometry the fluorescence intensity levels off at high values of the concentration without passing through a maximum provided self-absorption is negligible. ⟶

SAQ 3.3b (cont.)

(*iv*) The profile of the excitation band is likely to be more seriously affected than the emission band by the inner filter effect originating from the analyte itself.

(*v*) An inner filter effect at the emission wavelength of the analyte due to another compound in the solution can be confirmed by measuring the absorbance of the sample at this wavelength.

(*vi*) The presence of a compound having a weak absorption at the emission wavelength of the analyte in a buffer solution used in the preparation of the sample would seriously affect the results of an analysis.

(*vii*) If the absorbance of a sample at the excitation wavelength was greater than 0.2 the sample should be diluted $\times 10$.

Response

(*i*) True. At high concentrations the inner filter effect becomes more and more severe and I_f is no longer proportional to c but is proportional to e^{-c}.

(*ii*) False. The dependence on e^{-kcd} only gives rise to non-linearity. The inner filter effect not only gives rise to non-linearity, but at very high concentrations the intensity of radiation reaching the centre portions of the sample (upon which the emission monochromator is focussed) falls dramatically and hence the emission intensity detected falls to a corresponding extent. Hence a maximum in the calibration graph.

(*iii*) True. Unlike 90° geometry, with straight-through geometry the detector 'sees' all the fluorescence emitted by the sample. Once all the incident radiation is being absorbed the emission intensity remains constant and it does not matter whether it originates from the first 5 mm or the first 0.5 mm, it will be recorded (provided the inner filter effect at the excitation wavelength is the only factor affecting it).

(*iv*) True. The effect on the emission band is less serious unless the Stoke's shift is small and a wide band-pass is used.

(*v*) True. A better test is to run the absorption spectrum in this region. The analyte will not absorb in the region of its own emission spectrum. It would be more difficult to detect other absorbing species in the region of the excitation spectrum because of the absorption of the analyte itself.

(*vi*) False. The same quantity of the buffer solution is added to all samples and standards and so its effect would be constant. The absorption would cause some loss of sensitivity but if the absorption were weak this would not be significant (unless the concentration of the analyte were also very low). Note that exactly the same quantity of buffer would have to be added.

(*vii*) True. If the analyte absorbance is more than about 0.02 the concentration will be above the linear range. A high absorbance at the excitation wavelength will therefore indicate either that the concentration of the analyte is too high or that there is another component present which will cause an inner filter effect. In the first situation dilution of the sample will avoid problems, and even in the second case it may well improve matters provided the analyte concentration is well above the detection limit. If sensitivity cannot be sacrificed, however, dilution is no longer an option and it becomes necessary to find another remedy. If the absorbing species is known and its concentration can be easily determined (e.g. by making an absorbance reading in another region of the spectrum) it is possible to correct the fluorescence intensity and achieve acceptable results. Alternatively it may be possible to perform some chemical treatment on the absorbing species to remove it or change its absorption characteristics so that the interference is avoided.

****

SAQ 3.3c

Examine the excitation spectra of anthracene shown in Fig. 3.3g. What *three* pieces of evidence make it clear that there is a severe inner filter effect at a concentration of 10^{-5} mol dm^{-3}? Is it still present at 10^{-6} mol dm^{-3}?

The absorption spectrum of a 10^{-5} mol dm^{-3} solution is shown in Fig. 3.3h.

→

SAQ 3.3c (cont.)

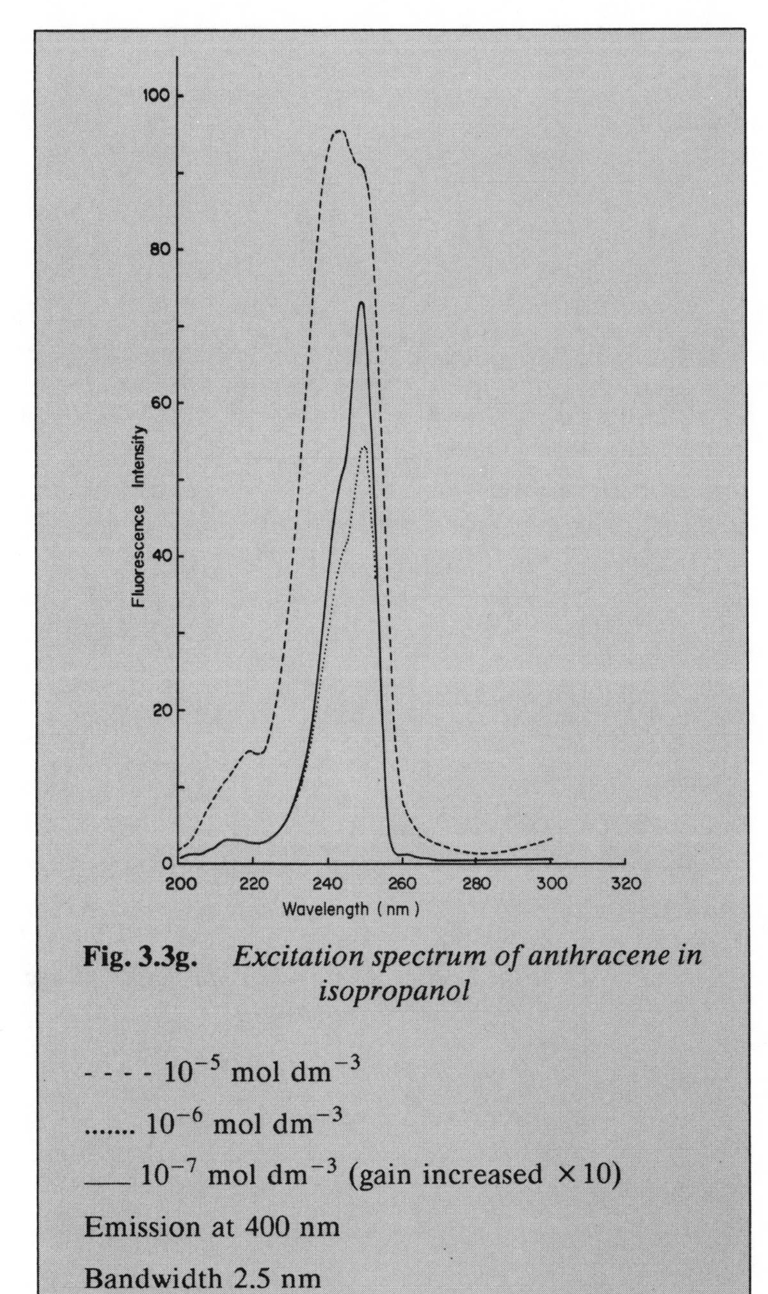

Fig. 3.3g. *Excitation spectrum of anthracene in isopropanol*

- - - - 10^{-5} mol dm^{-3}

.... 10^{-6} mol dm^{-3}

___ 10^{-7} mol dm^{-3} (gain increased $\times 10$)

Emission at 400 nm

Bandwidth 2.5 nm

⟶

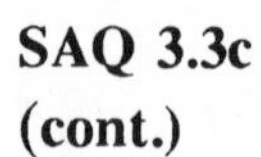

SAQ 3.3c (cont.)

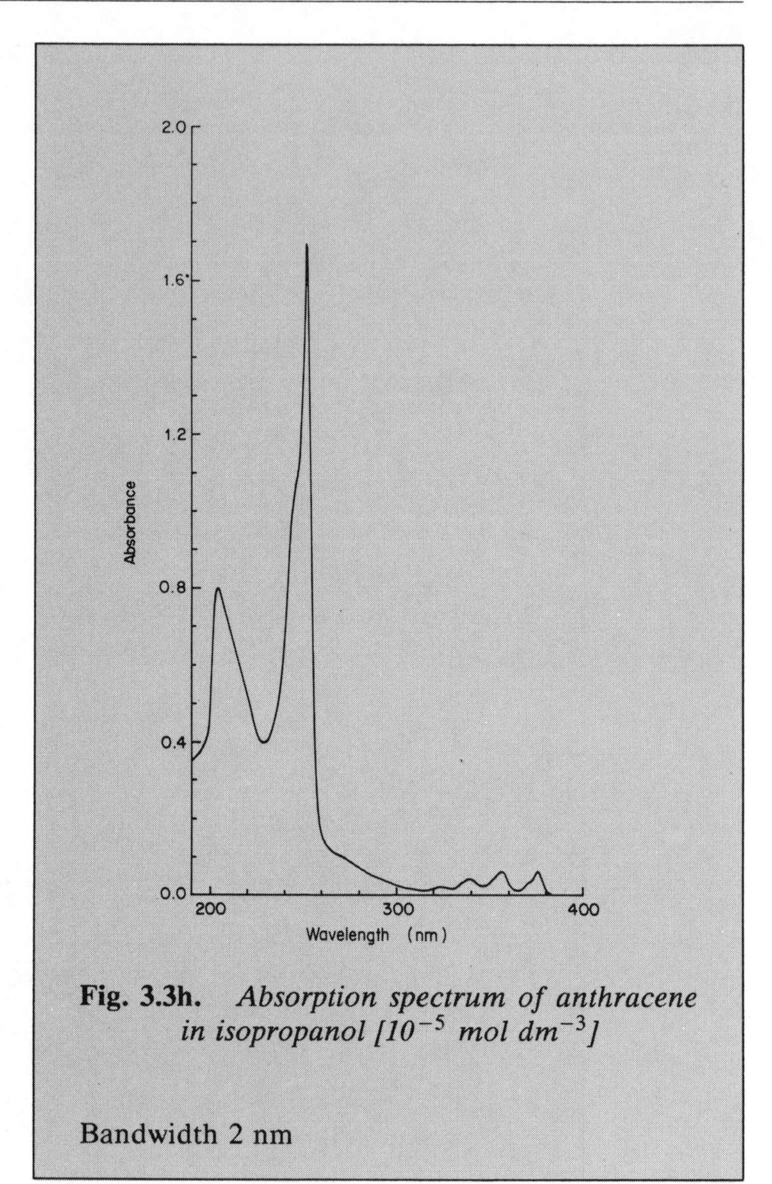

Fig. 3.3h. *Absorption spectrum of anthracene in isopropanol [10^{-5} mol dm^{-3}]*

Bandwidth 2 nm

Response

(*i*) The intensity of the band of the 10^{-5} mol dm^{-3} is only about 25% greater than that of the band of the 10^{-6} mol dm^{-3} solution. It should be 10 times greater.

(*ii*) The band maximum is at 243 nm at 10^{-5} mol dm^{-3} and 250 nm at 10^{-6} mol dm^{-3}, a shift of 7 nm.

(*iii*) The half-width of the band is about 23 nm at 10^{-5} mol dm^{-3} but only about 13 nm at 10^{-6} mol dm^{-3}. The band of the 10^{-6} mol dm^{-3} is at the same wavelength as the band of the 10^{-7} mol dm^{-3} solution and its half-width is much the same. It is not quite 10 times as intense however. This might be due to the inner filter effect but it could be due to the curvature caused by the higher terms in the expansion of $e^{-2.303\epsilon cd}$. The absorbance of a 10^{-6} mol dm^{-3} solution of anthracene based on Fig. 3.3h is 0.17 which is about 10 times the limit of linearity.

****

SAQ 3.4a

State whether each of the following statements is true or false:

(*i*) Fluorescence is quenched when a molecule in an excited electronic state is deactivated without the emission of a photon.

(*ii*) The fluorescence efficiency of a compound in solution is affected by the presence of other compounds.

(*iii*) Quenching can be used to measure the concentration of the quenching species.

⟶

SAQ 3.4a (cont.)

(*iv*) Self-quenching results in the intensity of fluorescence decreasing as the concentration increases.

Response

(*i*) True – a sound definition of quenching!

(*ii*) False – the fluorescence efficiency is an inherent property of the compound itself. The presence of impurities which are quenching agents will clearly affect the experimental determination of ϕ_f so great care must be taken to exclude them (and to remove dissolved oxygen). Since the fluorescence is always measured in solution the efficiency will always be subject to the influence of the solvent. Values of ϕ_f will therefore relate to a particular solvent which should always be stated (together with the temperature). Where possible the use of a non-polar hydrocarbon solvent such as *n*-hexane will give values least affected by the solvent.

(*iii*) True – fluorescence intensity falls progressively as the concentration of quenching agent rises in a reproducible though non-linear manner.

(*iv*) True – self-quenching (concentration quenching) is one of the processes which causes the gross departure of linearity of the calibration graph at high concentrations beyond that caused by the inner filter effect.

****

SAQ 3.4b

How could you demonstrate that the reduction of the intensity of the fluorescence of quinine sulphate at 440 nm by sodium chloride is due to quenching and not to the inner filter effect?

Response

Measure the absorbance of a sodium chloride solution at 440 nm. It will be zero (sodium chloride is colourless) so emission from the quinine is not being reabsorbed as would be required by an inner filter effect. This is a case of genuine quenching by the chloride ion.

****

SAQ 3.4c

Fill in the blanks in the following paragraph using the words listed below. Each word may be used once only.

Both quenching and the inner filter effect result in the .... of the intensity of fluorescence. They differ in that quenching does not involve .... of radiation while the inner filter effect does not involve .... state molecules. With quenching, excited state molecules transfer their .... to another molecule by .... or following the formation of a .... or by some other means of .... transfer. Other species present in the sample can give rise to the inner filter effect only if they .... radiation at the .... of excitation or .... The effect of quenching on an analysis can be corrected

⟶

SAQ 3.4c (cont.)

by adding an .... of the species responsible to all samples unless the .... of the analyte is near the .... limit.

energy	wavelength	complex
absorb	excited	detection
excess	reduction	emission
collision	absorption	concentration
radiationless		

Response

Both quenching and the inner filter effect result in the *reduction* of the intensity of fluorescence. They differ in that quenching does not involve *absorption* of radiation while the inner filter effect does not involve *excited* state molecules. With quenching, excited state molecules transfer their *energy* to another molecule by *collision* or following the formation of a *complex* or by some other means of *radiationless* transfer. Other species present in the sample can give rise to the inner filter effect only if they *absorb* radiation at the *wavelength* of excitation or *emission*. The effect of quenching on an analysis can be corrected by adding an *excess* of the species responsible to all samples unless the *concentration* of the analyte is near the *detection* limit.

****

SAQ 3.5a

Select the correct ending to the following sentence:

Radiation of wavelength 400 nm is less likely to cause photodecomposition of an organic molecule than radiation of wavelength 300 nm because

(*i*) its photon energy is higher.
(*ii*) it is visible radiation.
(*iii*) its photon energy is lower.
(*iv*) it is more readily absorbed.

Response

(*iii*) is correct.

Photon energy increases with frequency (Planck's Law, $\epsilon = h\nu$). The longer the wavelength the lower the frequency so (*i*) is wrong. (*ii*) is wrong because the fact that 400 nm is in the visible region (just) is not relevant. There is no distinction between uv and visible except that the human eye responds only to the latter. The ability to absorb radiation depends on the spacing between energy levels in the sample, not the wavelength of the radiation so (*iv*) is wrong.

****

SAQ 3.5b

Are the following statements true or false?

(*i*) There is no need to outgas a solution for fluorescence measurement because oxygen quenching occurs with the triplet state.

(*ii*) If the fluorescence of a solution decreases whilst it is being measured it is likely that photodecomposition is taking place

(*iii*) Fluorescence and photodecomposition are competitive processes because both result in the deactivation of excited state molecules.

(*iv*) Photodecomposition of the analyte always reduces the intensity of fluorescence.

Response

(*i*) False. Oxygen does quench fluorescence in some cases though it is admittedly more effective with phosphorescence. However, removal of oxygen is often necessary to avoid photodecomposition in a fluorescence method.

(*ii*) True. This is usually the first indication of photodecomposition.

(*iii*) False. At least, only half true! The processes are competitive but photodecomposition destroys the molecule rather than simply returning it to its ground state.

(*iv*) False. If one of the products of the photodecomposition is more fluorescent than the analyte the fluorescence intensity will increase.

****

SAQ 3.5c

Which of the following do *not* provide a possible cause of photodecomposition?

(*i*) The photon energy of uv radiation is comparable with the bond dissociation energy (per molecule) in many organic molecules.

(*ii*) Many compounds react with oxygen under strong uv irradiation.

(*iii*) Absorption of uv radiation can take a molecule into a high electronic excited state.

(*iv*) An excited molecule can lose its energy by collision with another molecule.

Response

(*iii*) and (*iv*).

(*iii*) – Absorption of uv radiation does not necessarily cause photodecomposition.

(*iv*) – When a molecule loses its excitation energy by collision it returns to the (stable) ground state.

Statements (*i*) and (*ii*) are valid causes of photodecomposition.

****

SAQ 3.5d

Which of the following steps would *reduce* the extent of photodecomposition?

(*i*) Increase the slit-width of the excitation monochromator.

(*ii*) Keep the excitation shutter closed until you are ready to take the reading.

(*iii*) Bubble oxygen-free nitrogen through the sample.

(*iv*) Reduce the slit-width of the emission monochromator.

(*v*) Use a shorter wavelength for excitation.

Response

(*ii*) and (*iii*).

(*i*) – Increasing the excitation slit-width increases the intensity of radiation falling on the sample and will therefore increase the extent of photodecomposition.

(*iv*) – Adjustments to the emission monochromator do not affect the conditions to which the sample is exposed.

(v) – Radiation of shorter wavelength has greater photon energy and so is more likely to cause photodecomposition.

****

SAQ 3.5e

The bond dissociation energy of the C—Br bond is 290 kJ mol^{-1}. The uv absorption spectrum of tribromomethane is shown in Fig. 3.5a. Determine whether the photon energy of radiation of wavelength

(i) 450 nm
(ii) 350 nm and
(iii) 250 nm

is sufficient to decompose the molecule and state whether decomposition will occur at each wavelength.

1 mole contains 6.0×10^{23} molecules
$h = 6.6 \times 10^{-34}$ J s
$c = 3.0 \times 10^{8}$ m s^{-1}

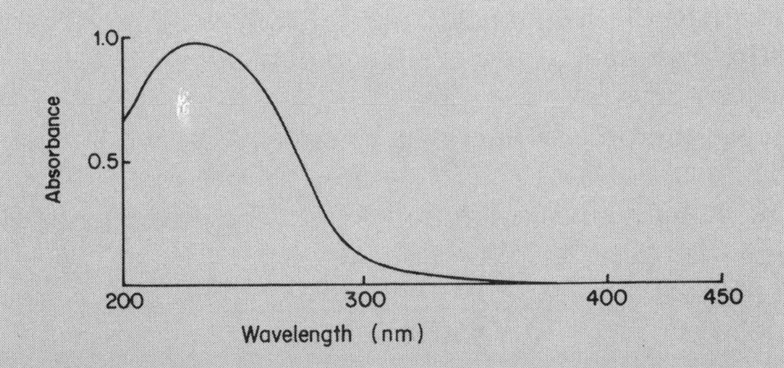

Fig. 3.5a. *Ultraviolet absorption spectrum of tribromomethane*

Response

(*i*) Photon energy is insufficient to cause decomposition.

(*ii*) Photon energy is now sufficient to decompose the molecule but no decomposition will in fact occur because the compound does not absorb at 350 nm.

(*iii*) Photon energy is more than enough to cause decomposition and the radiation is strongly absorbed so the compound will decompose.

Since the absorption of radiation is the result of an interaction between one molecule and one photon we have to compare the energy required to break one C—Br bond with the energy of one photon. The value quoted for the bond dissociation energy refers to 1 mole, ie 6.0×10^{23} C—Br bonds. The energy to break one bond is therefore

$$(290 \times 1000)/(6.0 \times 10^{23})\ \text{J} = 4.8 \times 10^{-19}\ \text{J}.$$

The value of the photon energy is obtained by using Planck's Law, $\epsilon = h\nu$, which we referred to in SAQ 3.5a. We actually need the frequency, ν, to apply this relation and, unfortunately we were provided with the wavelength. The fact that you were also given the value of c, the velocity of light should have reminded you that to calculate the frequency from the wavelength, λ, we need another little fundamental law of nature $c = \nu\lambda$.

Hence the photon energy is given by $\epsilon = hc/\lambda$.

The actual values corresponding to the wavelengths given are:

$$\text{at 450 nm} = \frac{6.6 \times 10^{-34} \times 3.0 \times 10^{8}}{450 \times 10^{-9}}$$

$$= 4.4 \times 10^{-19}\ \text{J}$$

$$\text{at 350 nm} = \frac{6.6 \times 10^{-34} \times 3.0 \times 10^{8}}{350 \times 10^{-9}}$$

$$= 5.7 \times 10^{-19} \text{ J}$$

at 250 nm $= \dfrac{6.6 \times 10^{-34} \times 3.0 \times 10^{8}}{250 \times 10^{-9}}$

$$= 7.9 \times 10^{-19} \text{ J}$$

****

SAQ 3.6a

Are the following statements true or false?

(*i*) The absolute sensitivity of a particular compound is given by the slope of the calibration graph.

(*ii*) The instrumental sensitivity stated as '0.005 ng dm^{-3} of quinine' means that the instrument is capable of recording that concentration of quinine with a signal/noise ratio of 50.

(*iii*) The instrumental sensitivity is now commonly expressed in terms of the signal/noise ratio of the Raman band of water excited at 350 nm with a band-pass of 10 nm.

(*iv*) The method sensitivity is generally governed by the emission from the blank.

Response

(*i*) False. The slope is not only governed by the spectral characteristics of the compound, but by instrumental factors also.

(*ii*) False. This method of expressing instrumental sensitivity gives the detection limit for quinine for the instrument. This is the concentration at which the signal/noise ratio is 2, not 50.

(*iii*) True – in every respect!

(*iv*) True. The only occasions where other factors might be of greater importance are methods where solvent extraction is used as a pre-concentration technique. This increases the concentration of the sample presented to the instrument but it often reduces the blank too.

****

SAQ 3.6b

Identify the item which is out of place (in the context of fluorescence sensitivity) in each of the following groups:

(*i*) Incident radiation intensity, fluorescence efficiency, detector sensitivity, monochromator band-pass.

(*ii*) Incident intensity, detector sensitivity, monochromator transmittance, monochromator aperture.

(*iii*) Raman scattering, Rayleigh scattering, stray light, Tyndall scattering.

(*iv*) Buffer solution, fluorescence reagent, organic solvent, colloidal solution.

(*v*) Distillation, membrane filtration, liquid chromatography, charcoal treatment.

⟶

SAQ 3.6b (cont.)

> (*vi*) Change wavelength of excitation, use frontal illumination, increase band-pass of emission monochromator, change solvent.

Response

(*i*) Fluorescence efficiency. This is a property of a fluorescent molecule. The other items are all factors involved in instrumental sensitivity.

(*ii*) Monochromator aperture. All the other items are wavelength dependent. The aperture is determined by the size of the optical components.

(*iii*) Raman scattering. All the other items have the same wavelength as the exciting radiation. (Stray light – because it's not scattering, or named after a famous scientist – is correct but trivial!)

(*iv*) Colloidal solution. All the other items are possible sources of fluorescent impurities. Colloidal solutions give rise to Tyndall scattering.

(*v*) Membrane filtration. This removes suspended particles but not fluorescent impurities. The other items can remove both.

(*vi*) Use frontal illumination. This is used with turbid or absorbing samples. The other items are methods of discriminating between fluorescence and Raman scattering.

****

SAQ 3.6c

Rearrange List B so that each item becomes a remedy for the corresponding problem encountered in sample preparation given in List A.

List A	*List B*
Metal impurities in fluorescence reagents	Redistillation
Bubbles of gas	Filtration
Fluorescent impurities in solvents	Charcoal treatment
Fluorescent impurities in a buffer solution	Heat the sample
Suspended particles	Ion exchange

Response

List A	*List B*
Metal impurities in fluorescence reagents	Ion exchange
Bubbles of gas	Heat the sample
Fluorescent impurities in solvents	Redistillation
Fluorescent impurities in a buffer solution	Charcoal treatment
Suspended particles	Filtration

****

SAQ 3.6d

Rearrange List B so that each item becomes a remedy for the corresponding sampling problem in list A.

List A	*List B*
High blank	Increase emission slit-width
Raman interference	Use auxiliary emission cut off filter
High stray light	Use frontal illumination
Strong inner filter effect.	Use zero suppression

Response

List A	*List B*
High blank	Use zero suppression
Raman interference	Increase emission slit-width
High stray light	Use auxiliary emission filter
Strong inner filter effect	Use frontal illumination

****

SAQ 3.6e

> The main infrared absorption band of water occurs at 3300 cm^{-1}. Calculate the wavelength of the Raman band when an aqueous sample is exposed to exciting radiation at 390 nm during a fluorescence experiment.

Response

448 nm.

The wavenumber of the Raman band is $\dfrac{\nu_o}{c} - \dfrac{\nu_{vib}}{c}$,

where

(*i*) ν_o/c is the wavenumber of the exciting radiation which is the reciprocal of its wavelength, $1/\lambda_{ex}$. The wavelength is 390 nm or 390×10^{-7} cm so the wavenumber is

$$\frac{1}{390 \times 10^{-7}}$$

$$= \frac{1000 \times 10^4}{390}$$

$$= 2.56 \times 10^4$$

$$= 25\,600 \text{ cm}^{-1}$$

(*ii*) ν_{vib}/c is the wavenumber of the ir absorption band, 3300 cm^{-1}.

Hence $\nu_o/c - \nu_{vib}/c$ is $25\,600 - 3300 = 22\,300$ cm^{-1}.

Radiation of wavenumber 22 300 cm^{-1} has a wavelength of

$$\frac{1}{22\,300}\ \text{cm} = \frac{10\,000\,000}{22\,300}\ \text{nm}$$

$$= 448\ \text{nm}.$$

****

SAQ 3.6f

Emission spectra of a solution of anthracene in propan-2-ol (10^{-7} mol dm^{-3}) recorded with the emission slit width set to 2.5, 5, 10 and 20 nm are shown in Fig. 3.6a. The excitation slit-width was 2.5 nm throughout and the excitation wavelength was 354 nm. Raman spectra of propan-2-ol recorded under the same conditions are shown in Fig. 3.6b. The ordinate expansion factor was adjusted for each spectrum to maintain the peak intensities approximately constant. This is recorded on the spectra.

(*i*) How does the Raman spectrum interfere with the fluorescence band?

(*ii*) Measure the peak height of the Raman band and the strongest fluorescence peak in each spectrum. Divide all these readings by the ordinate expansion factor (f), printed on each spectrum, to bring them to the same intensity scale. Does increasing the slit-width improve discrimination in favour of the fluorescence peak?.

⟶

SAQ 3.6f (cont.)

(*iii*) Refer to the excitation spectrum of anthracene, Fig. 3.6c. Suggest an alternative excitation wavelength to remove the Raman interference. How much loss in intensity would result from this change?

(*iv*) Check whether the use of carbon tetrachloride with Raman bands at shifts of 459 and 776 cm^{-1} would be a more effective way of removing the interference.

(*v*) Which ir band of propan-2-ol does the strongest Raman band correspond to?

Response

(*i*) The Raman band coincides with the strongest fluorescence band at 400 nm.

(*ii*) The measurements are as follows:

Slit-width/nm	2.5	5	10	20
Raman peak height	32	43	38	31
Expansion for Raman	50	20	5	2
Raman peak/expansion	0.64	2.15	7.6	5.5
Fluorescence peak height	54.5	50	42	51
Expansion for Fluorescence	20	5	1	0.5
Fluorescence/expansion	2.72	10.0	42.0	102
Fluorescence/Raman	3.35	3.75	4.85	6.18

Doubling the slit-width increases the fluorescence intensity by a factor of about 4 as expected for broad bands. The corresponding increase in the Raman intensity is a factor of about 3 up to a slit-width of 10 nm and only $\times 2$ from 10 to 20 nm. This is due to the narrower width of the Raman band. Hence increasing the slit-width does enhance the fluorescence relative to the Raman band. A more accurate calculation in which the bands are corrected for background emission, and the fluorescence intensity is corrected for Raman interference gives the values for the ratio of fluorescence intensity to Raman intensity quoted in the bottom line of the above table.

These show a gain of a factor of about 2 with 20 nm slits compared with 2.5 nm slits.

(*iii*) The excitation peak at 336 nm could be used. This appears to be only about 2/3 the intensity of 354 nm but the latter is also enhanced by Raman interference. (The separation of 354 and 400 nm is equal to the Raman shift of the solvent so interference occurs in both spectra.) 373 nm could also be used since it has much the same intensity as 336 nm. However, this is the 0,0 band which overlaps the 0,0 band of the emission spectrum so we cannot record this band because of the presence of the Rayleigh peak. Because the bands in the anthracene spectrum are almost equally spaced, whichever excitation wavelength we use the Raman band will overlap one of the emission bands as shown in Fig. 3.6d. The true emission spectrum is shown in Fig. 3.6e where the solution is excited at 250 nm – much too far away for Raman interference to occur.

(*iv*) The longest wavelength band in the Raman spectrum of carbon tetrachloride excited at 354 nm is 364 nm $= 10^7/(28\,249 - 776)$. This is well clear of the emission spectrum of anthracene and so a change to this solvent would indeed be the best way of avoiding Raman interference.

(*v*) The Raman shift is 3000 cm^{-1} so it corresponds to the C—H stretching band in the ir spectrum (not the O—H as you might have expected). The Raman band is actually at 397 nm which corresponds to a wavenumber of 25 200 cm^{-1}. The wavenumber of the exciting radiation is $10^7/354 = 28\,200$ cm^{-1}. Hence the Raman shift is 28 200 − 25 200 cm^{-1} from 354 nm.

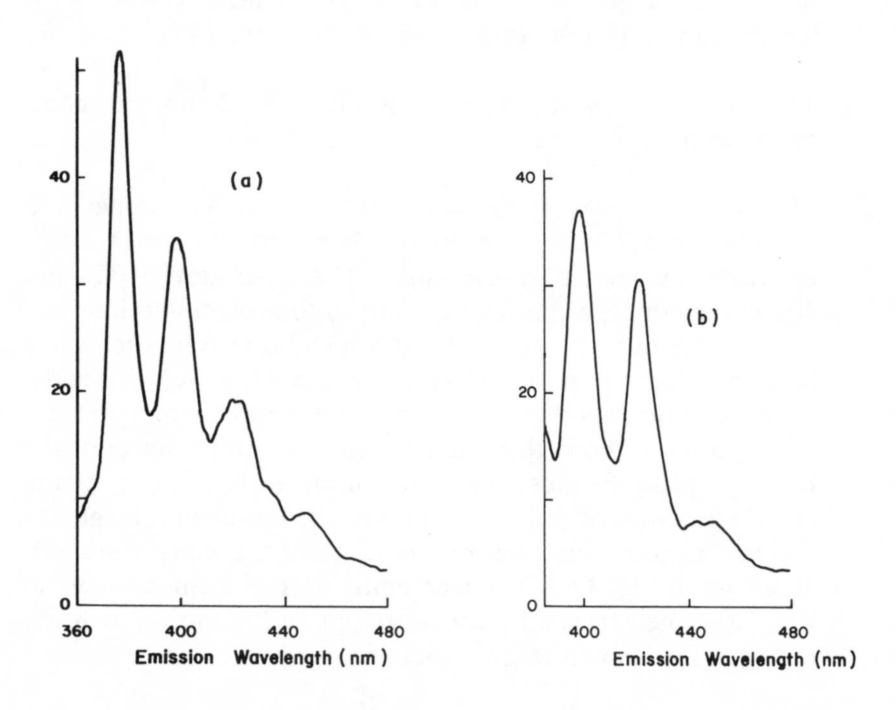

Fig. 3.6d. *Fluorescence emission spectra of anthracene in propan-2-ol (10^{-7} mol dm^{-3})*

(*a*) Excitation at 336 nm
(*b*) Excitation at 373 nm

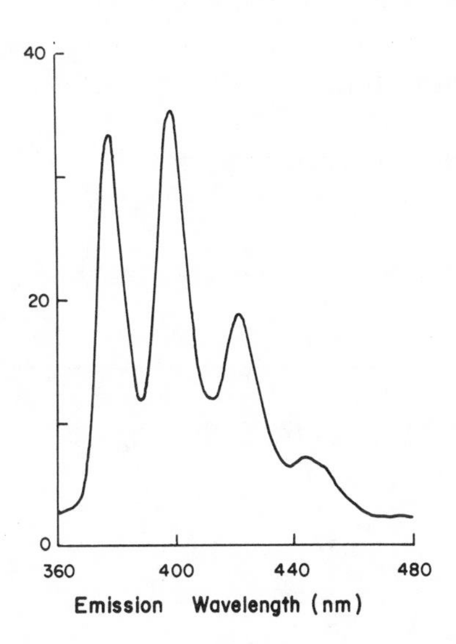

Fig. 3.6e. *Fluorescence emission spectrum of anthracene in propan-2-ol (10^{-7} mol dm^{-3}) [Excitation at 250 nm]*

****

SAQ 3.6g

> Draw up a list of all the factors which contribute to the blank in a fluorescence assay at low concentrations.

Response

(*i*) Suspended particles.

(*ii*) Other fluorescent impurities present in the sample.

(*iii*) Fluorescent impurities present in solvents and reagents.

(*iv*) Rayleigh scattering.

(*v*) Raman scattering.

(*vi*) Luminescent impurities in cells.

****

SAQ 4.2a

Which of the following lanthanides is fluorescent in solution?

Eu Lu Ce Tb

Response

Eu and Tb.

****

SAQ 4.2b

(*a*) What simple uranium-containing species is fluorescent?

(*b*) What colour is the fluorescence?

(*c*) Under what conditions is it enhanced?

(*d*) Why does its emission band show fine structure?

Response

(*i*) UO_2^{2+}

(*ii*) Green.

(*iii*) In the presence of sulphuric acid.

(*iv*) This is vibrational fine structure of a molecular species.

****

SAQ 4.2c

Which of the following compounds could be analysed using their natural fluorescence?

tryptophan
vitamin A
ethanedicarboxylic acid
cortisol

Response

Vitamin A and cortisol.

Vitamin A has a chain of five conjugated double bonds. Cortisol is a large polycyclic steroid molecule which fluoresces in conc. sulphuric acid solution.

Tryptophan fluoresces at too short a wavelength to be useful.

Carboxylic acids are not naturally fluorescent. This one would condense with resorcinol to give a fluorescent derivative.

****

SAQ 4.3a

Draw the structure of the complex formed between boric acid and 1,2-dihydroxybenzene. Why should it be fluorescent?

Response

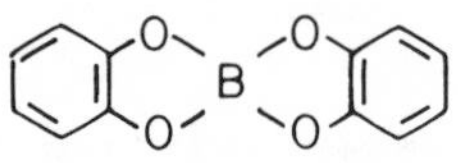

It forms a cyclic structure like the benzoin complex.

****

SAQ 4.3b

Draw the structure of the complex formed between tin and 3-hydroxyflavone assuming that it is octahedral (see Fig. 4.3a).

Response

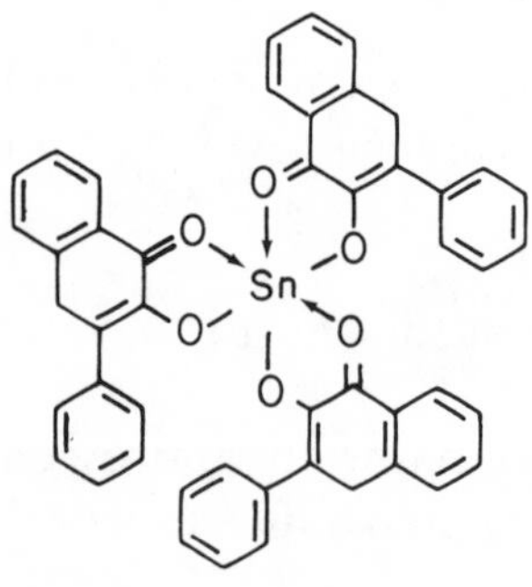

****

SAQ 4.3c

Are the following statements about complexes of the lanthanides true or false?

(*i*) The excited state is more easily deactivated than in transition metal complexes.

(*ii*) The emission spectrum is a mirror image of the excitation spectrum.

(*iii*) The emission spectrum consists of a number of very sharp bands.

(*iv*) The lifetime of the emission is greater than 1 μs.

(*v*) The emission can be described as phosphorescence.

(*vi*) The emission can be observed with the sample in the liquid state.

Response

(*i*) False. The 4f electrons of the lanthanides are buried too deep inside the core to be easily affected by external influences. d-electrons in transitional metals are far more accessible.

(*ii*) False. The absorption occurs through the organic part of the molecule via the singlet state. The emission occurs from the 'atomic' states involving the 4f electrons following internal conversion.

(*iii*) True. Sharp bands or lines are characteristic of the 'atomic' nature of this emission.

(*iv*) True. Again a characteristic property of these complexes.

(*v*) False. Phosphorescence involves emission from a triplet state.

(*vi*) True. Screening of the 4f levels by the valence electrons inhibits de-activation of the excited state molecules even in the liquid phase so it is not necessary to freeze the sample.

****

SAQ 4.3d

State whether completion of the following sentence by the phrases (*i*) to (*v*) gives a statement which is true or false:

Derivatisation of organic compounds

(*i*) shifts any natural fluorescence to longer wavelength.

(*ii*) increases the size of the π-system.

(*iii*) involves elimination of small molecules like H_2O and HCl.

(*iv*) gives a product with a higher fluorescence efficiency.

(*v*) is specific because a reagent reacts only with one class of compound.

Response

All the statements are true! Together they provide a summary of the principles underlying organic derivatisation.

The emission (and absorption) shifts to longer wavelengths because the size of the π-system is increased by the elimination of small molecules when the reagent and the analyte condense to form an additional double bond. This frequently increases the rigidity of the molecule, especially when another ring is formed and so increases ϕ_f. In other cases a lone pair may be delocalised and so remove a low-lying $n \rightarrow \pi^*$ excited state. Reagents usually react with analytes having a particular functional group and, in many cases, only when the reacting groups can take up the correct relative positions – in other words, the reaction is controlled by a steric factor.

****

SAQ 4.3e

State *three* ways in which the emission from a lanthanide complex differs from phosphorescence.

Response

(*i*) The transitions concerned involve non-bonding electrons in atomic f-orbitals, not a triplet state of the complex molecule.

(*ii*) Several sharp bands are observed in the spectrum.

(*iii*) The emission is observed from a liquid sample without special treatment.

****

SAQ 4.3f

List the main factors which can be used to achieve higher selectivity in a fluorimetric analysis.

Response

(*i*) Choice of complexing or derivatising agent, in particular to take advantage of steric factors which may restrict the range of analytes with which the reagent will react.

(*ii*) Control of pH and other conditions under which the reaction takes place in order to favour a particular analyte.

(*iii*) Use of solvent extraction under appropriate conditions to favour isolation of a particular species.

(*iv*) Use of time resolution where the analyte has a suitable luminescence lifetime.

****

SAQ 4.4a

The fluorescence of quinine is quenched by the presence of chloride ions, and the following data obtained.

$[Cl^-]$/mol dm^{-3}	Fluorescence Intensity, I_f
0.00	100 $[I_f = I_f^o]$
0.01	30.05
0.02	20.7
0.03	14.7
0.04	11.0
0.05	8.2
0.06	6.7
0.07	5.9
0.08	5.4

Plot the data using the Stern–Volmer equation. Do you get a straight line?

Response

You should get a reasonable straight-line plot, at least for concentrations above 0.02 mol dm^{-3}. My plot and the calculated data are as follows:

$[Cl^-]$/mol dm^{-3}	I_f	I_f^o / I_f	$I_f^o / I_f - 1$
0.01	30.5	3.28	2.28
0.02	20.7	4.83	3.83
0.03	14.7	6.80	5.80
0.04	11.0	9.09	8.09
0.05	8.2	12.20	11.20
0.06	6.7	14.90	13.90
0.07	5.9	17.00	16.00
0.08	5.4	18.50	17.50

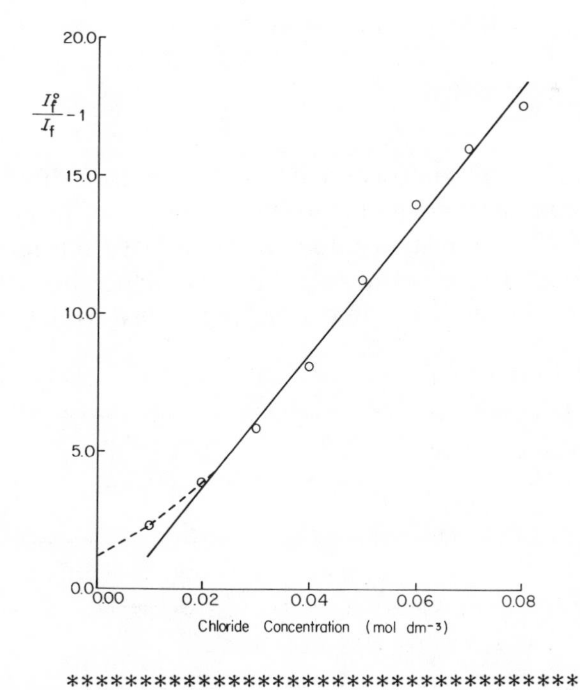

****

SAQ 4.4b

Select the phrase, (*i*) to (*iv*), which correctly completes the following sentence:

Chloride ions cannot be determined in the presence of bromide ions by the quenching of quinine fluorescence because

(*i*) the lifetime of the excited quinine molecule is too short.

(*ii*) the sensitivity increases with the mass of the ion.

(*iii*) quenching techniques are not specific.

(*iv*) the calibration graph is non-linear.

Response

(*iii*) is correct.

The lifetime of the excited species affects the sensitivity for both ions ion a similar way and does not provide a means of discriminating between them. The sensitivity does increase with the mass of the ion but this does not explain why the ions cannot be determined separately. Non-linearity of the calibration is also irrelevant.

****

SAQ 4.4c

Select the correct answer:

To obtain 50% quenching with 0.1% of oxygen would require a fluorescer having a fluorescence lifetime of

(*i*) 10 μs
(*ii*) 0.1 μs
(*iii*) 10 ms
(*iv*) 0.1 ms

Response

(*i*) is correct.

The sensitivity of quenching methods is proportional to K, the slope of the Stern–Volmer plot which itself is proportional to the lifetime of the excited species. We have already seen that with pyrene, $\tau_0 = 10^{-6}$ s, 50% quenching was achieved with 1% oxygen. To get 50% quenching with 0.1% oxygen we need to increase the value of τ_0 by a factor of 10, so a fluorescer of lifetime 10^{-5} s = 10 μs is required.

****

SAQ 4.5a

Fig. 4.5d shows the normal emission and the synchronous scan spectra of anthracene. What are the three chief differences between these two spectra?

→

SAQ 4.5a (cont.)

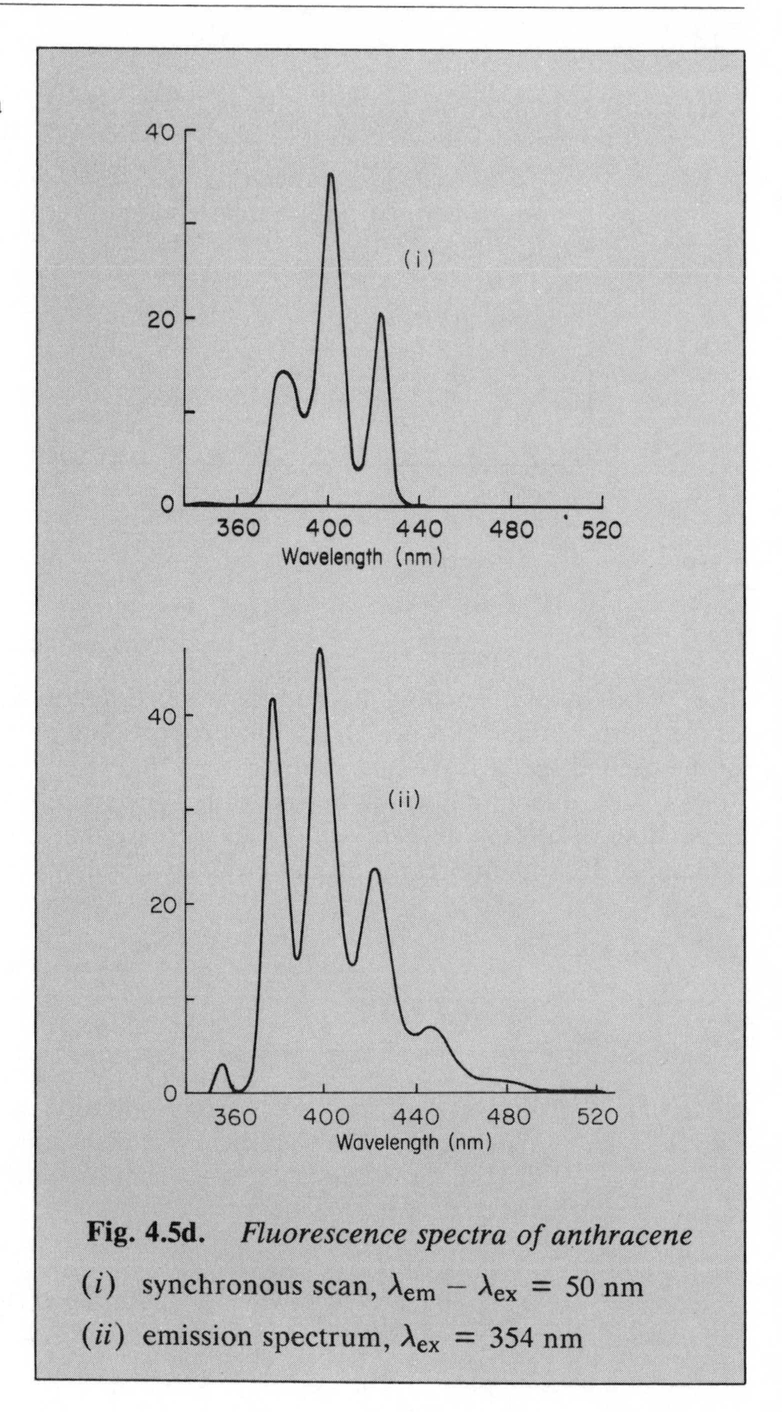

Fig. 4.5d. *Fluorescence spectra of anthracene*

(*i*) synchronous scan, $\lambda_{em} - \lambda_{ex} = 50$ nm

(*ii*) emission spectrum, $\lambda_{ex} = 354$ nm

Response

— The band has an overall width of about 60 nm in the synchronous scan spectrum and about 90 nm in the emission spectrum.

— The 468 nm peak and other long wavelength features are absent in the synchronous scan spectrum.

— There is no Rayleigh peak in the synchronous scan spectrum (354 nm in the normal emission spectrum).

****

SAQ 4.5b

What advantages do the following two differences between an emission spectrum and a synchronous scan spectrum confer on the synchronous scan spectrum?

(*i*) The band-width of a synchronous scan spectrum is less than that of an emission spectrum.

(*ii*) There is no Rayleigh peak in synchronous scan spectrum.

Response

(*i*) The narrower band-width reduces the problems of overlap between the bands of the components of a mixture.

(*ii*) The absence of a Rayleigh band avoids blotting out a region of the spectrum and, if the level of direct scattering is high, there is no danger of blinding the photomultiplier. A 'dirty' sample with a high direct scatter will contribute a fairly constant background to the spectrum which can be off-set.

****

SAQ 4.5c

Fig. 4.5e shows fluorescence spectra of a mixture of fluorene, naphthalene and anthracene recorded under three different conditions. In what ways do these spectra demonstrate the advantages of the synchronous scanning technique when examining a mixture of fluorescent compounds?

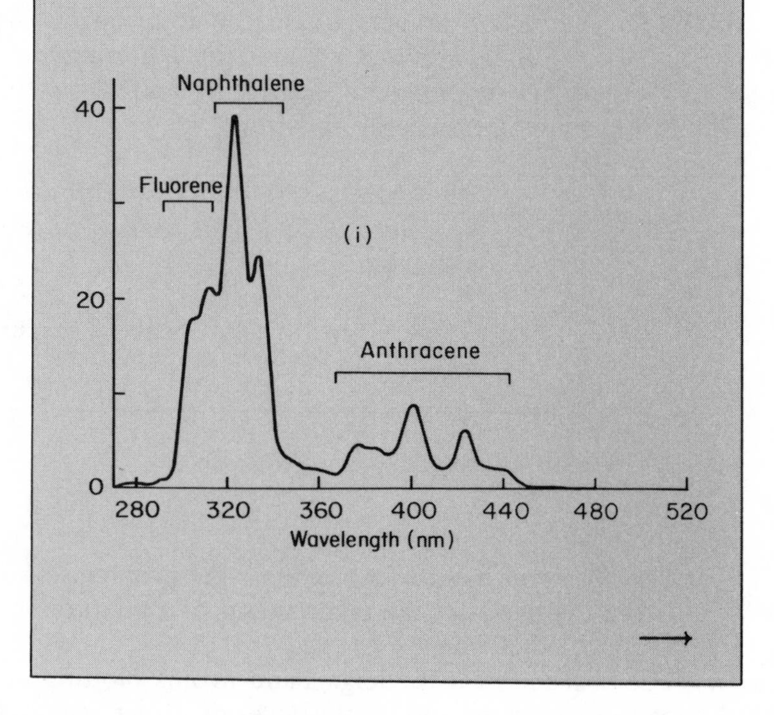

→

SAQ 4.5c (cont.)

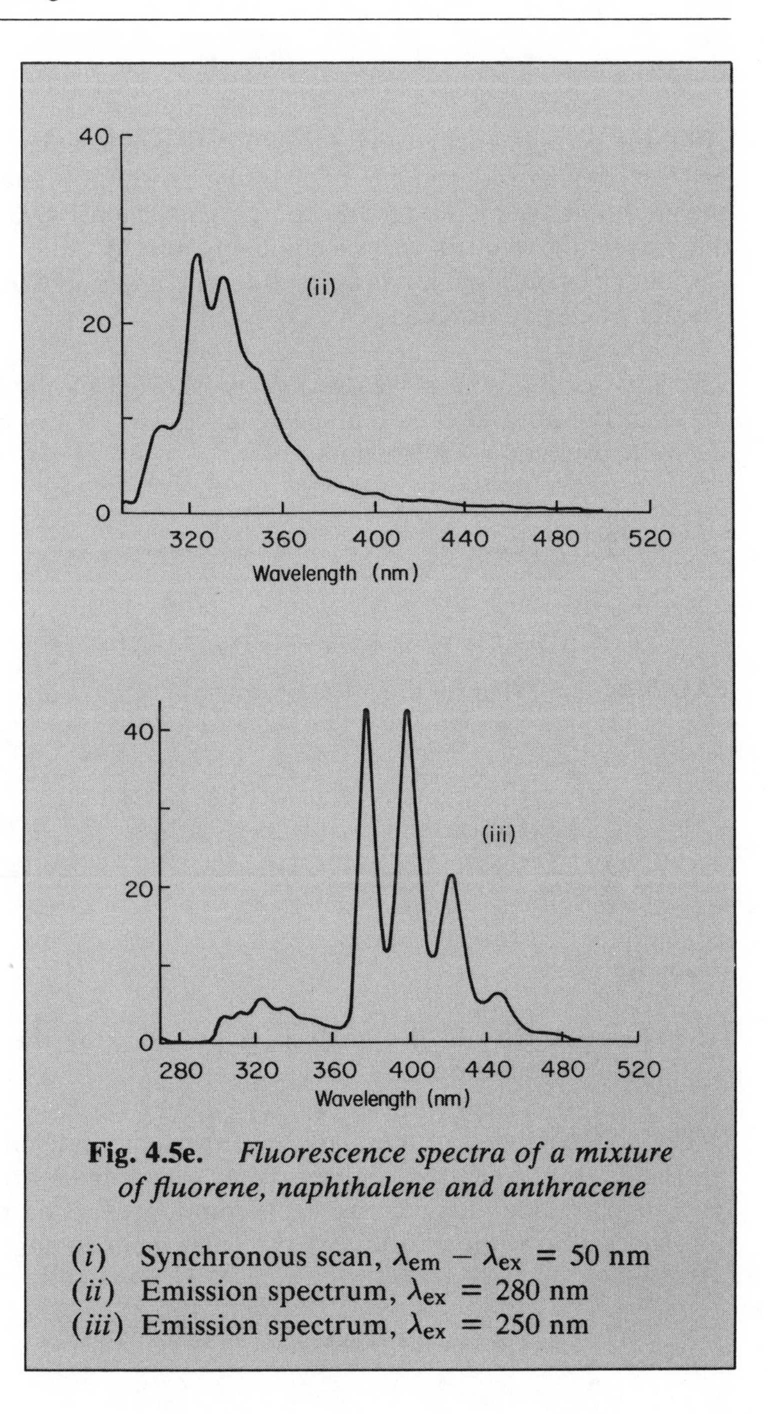

Fig. 4.5e. *Fluorescence spectra of a mixture of fluorene, naphthalene and anthracene*

(*i*) Synchronous scan, $\lambda_{em} - \lambda_{ex} = 50$ nm
(*ii*) Emission spectrum, $\lambda_{ex} = 280$ nm
(*iii*) Emission spectrum, $\lambda_{ex} = 250$ nm

Response

All three components give rise to separate bands which can clearly be seen (though with some overlap in the case of fluorene and naphthalene) and are of reasonably comparable intensity. Excitation at 280 nm would result in anthracene being missed while excitation at 250 nm gives only weak bands for fluorene and naphthalene which are not as clearly resolved.

The band of fluorene is also seriously distorted by the inner filter effect in the normal emission spectrum because it lies only 30 nm from the excitation wavelength.

****

SAQ 5.2a

Why is the lack of specificity in a derivatising agent

(*i*) no problem in either tlc or hplc,
(*ii*) often an advantage in hplc?

Response

(*i*) The selectivity in the analysis is provided by the chromatographic separation.

(*ii*) All the analytes are derivatised at the same time and, in many cases, the excitation and emission wavelengths are the same for each one. This allows each component to be detected in succession without adjusting the instrument settings.

****

SAQ 5.2b

When using a tlc plate scanning accessory to locate fluorescent spots on a tlc plate, which of the following settings would you use for:

(*i*) the excitation monochromator;
(*ii*) the emission monochromator?

— wide slit-width
— narrow slit-width
— short wavelength
— long wavelength
— direct light setting
— cut-off filter to exclude short wavelength

Response

(*i*) wide slit-width; short wavelength.

If you don't know what compounds may be present you will not know the precise excitation wavelength and the choice of a short wavelength is more likely to excite fluorescence. A wide slit-width provides a wider range of exciting wavelengths so that there is a greater chance of including radiation within the excitation bands of the compounds present. It also increases the intensity of the emission. The direct light setting would give rise to scattering at all wavelengths and obscure the fluorescence from the spots. A low wavelength cut-off filter would exclude the very wavelengths we need to excite the fluorescence.

(*ii*) direct light setting; cut-off filter to exclude short wavelength

The direct light setting would enable us to detect the fluorescence whatever its wavelength. If this were not available it would be advisable to use the greatest possible slit-width and set the emission wavelength about 70 nm higher than the excitation wavelength. The cut-off filter would cut down the scattered exciting radiation and reduce the background level this would cause as stray light in the emission monochromator.

****

SAQ 5.2c

State whether each of the following effects resulting from the formation of a fluorescent derivative is:

(*A*) associated with pre- or post-column derivatisation;
(*B*) an advantage or a disadvantage in an analysis.

(*i*) Absorption of the eluate is increased.
(*ii*) Detection of the eluted components is delayed.
(*iii*) The efficiency of the separation is affected.
(*iv*) All the analytes are derivatised simultaneously.
(*v*) Conditions for the optimum separation of the analytes are not affected.

Response

(*i*) Post-, disadvantage. High absorption in the eluate can cause loss of sensitivity and other problems due to the inner filter effect.

(*ii*) Post-, disadvantage. This is due to increasing the path-length between column and detector which impairs the resolution because the separated components can spread out in the mobile phase by diffusion.

(*iii*) Pre-, disadvantage. In the end it may well be possible to restore the efficiency for the derivatised compounds by re-optimising the conditions but this would take time.

(*iv*) Pre-, advantage. It is much easier to derivatise all the analytes before injection on to the column, especially when the reactions are at all complicated.

(*v*) Post-, advantage. This is perhaps the only advantage of post-column derivatisation and is the converse of (*iii*).

****

SAQ 5.2d

Draw the structure of the dansyl chloride molecule. Why is it a satisfactory derivatising agent for fluorescence detection in hplc?

Response

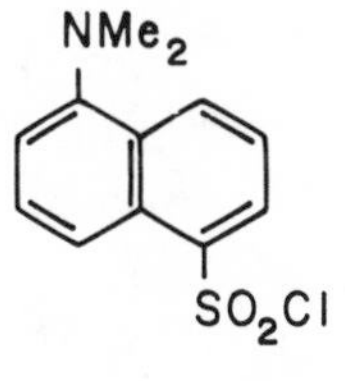

NMe_2

SO_2Cl

The reactive chlorine atom enables the reagent to be linked to the analyte species relatively easily by loss of HCl with a group containing a replaceable hydrogen atom (NH_2, OH etc.). The molecule also provides a large chromophore with a high absorptivity and rigid structure. The derivatives are readily separated and so pre-column derivatisation can be used.

****

SAQ 5.3a

What are the chief advantages and disadvantages of each of the four fluorescent labels given below?

Label	λ_{ex}/nm	λ_{em}/nm	ϕ_f	ϵ/dm^3 mol^{-1} cm^{-1}	Decay Time/ns
FITC	492	518	0.68	72 000	4.5
RBITC	550	585	0.30	12 300	3.0
DNS	340	500	0.30	3 400	14.0
Fluoram	394	475	0.10	6 300	7.0

Response

(*i*) FITC — advantages: highest value of ϵ and ϕ_f, high sensitivity absorption and emission well up in visible.

disadvantage: very small Stokes' shift (26 nm)

(*ii*) RBITC — advantages: high value of ϕ_f and high value of ϵ, highest wavelengths of absorption and emission.

disadvantage: low Stokes' shift (35 nm)

(*iii*) DNS — advantage: highest Stokes' shift (>100 nm)

disadvantages: low ϵ and ϕ_f values excitation wavelength rather low.

(*iv*)	Fluoram	advantages	good Stokes shift, ϵ value better than DNS, excitation wavelength higher than DNS.
		disadvantage	lowest value of fluorescence efficiency.

****

SAQ 5.3b

Write the equation for the antibody (Ab) – antigen reaction in conditions where there is an excess of antigen, labelled (Ag*) and unlabelled (Ag), competing for sites on the antibody. What is the analyte in this system?

Response

$$2Ab + 2Ag + 2Ag^* = (Ab–Ag) + (Ab–Ag^*) + Ag + Ag^*$$

Ag is the analyte.

****

SAQ 5.3c

In what respects does FIA score over RIA? (Four points were noted in the text.)

Response

— Lower overall cost.
— No health hazard from radioisotopes.
— Longer shelf-life of reagents.
— Homogeneous assay possible.

****

SAQ 5.3d

Explain why homogeneous assay is possible with FIA but not with RIA.

Response

The fluorescent properties (λ_{ex}, λ_{em} and ϕ_f) of the labelled antigen may differ for the Ab–Ag* complex and the free antigen. Hence we may be able to determine the complex and the free antigen at different wavelengths, or, if only the fluorescence efficiency differs, the total fluorescence intensity can be related to the concentration of either species. With RIA the activity of the labelled antigen is unaffected by complexing.

****

SAQ 5.3e

What reagents and solutions would you expect to find in a fluoroimmunoassay kit?

Response

(*i*) The antiserum containing the antibody raised against the analyte (antigen).

(*ii*) The labelled antigen carrying a fluorescent label.

(*iii*) Buffer solutions and other reagents used in the assay.

In methods involving enzyme hydrolysis of the Ab–Ag complex to release the fluorescent label, a solution of the appropriate enzyme would also be included.

****

SAQ 5.3f

Are the following statements true or false?

(*i*) The chief attraction of the immunoassay technique is its high sensitivity.

(*ii*) Labelling techniques in immunoassays require extremely sensitive methods of detection because of the very low molar concentration of the labelled species.

(*iii*) Anti-oxytocin is the antigen present in antiserum raised in order to determine oxytocin by FIA.

(*iv*) Radioactive labels are are more likely to affect the immune reaction than fluorescent labels.

⟶

SAQ 5.3f (cont.)

(*v*) Small analyte molecules may require linking to a protein molecule in order to provoke an immune response in a laboratory animal.

(*vi*) In a fluorescence enhancement method the free labelled antigen has a higher fluorescence efficiency than the bound antigen.

Response

(*i*) False. It is the selectivity which is the chief attraction. In terms of the analyte, the method is no more sensitive than direct fluorescence measurement.

(*ii*) True. The high molar mass of antibodies and antigens limits their molar concentration – even though the concentration may seem quite high in w/v terms.

(*iii*) False. Anti-oxytocin is the antibody which will react with the analyte, oxytocin itself.

(*iv*) False. Radioactive isotopes have no 'chemical' effect on the bonding properties of antibodies or antigens. Fluorescent labels are quite large molecular groups attached to the antigen molecule which could well cause steric effects if close to a binding site.

(*v*) True. In immunoassay jargon these are referred to as 'haptens'.

(*vi*) False. The complex must have a higher fluorescence efficiency than the free antigen if the fluorescence is to be enhanced.

****

SAQ 5.4a

Identify the phrase(s) which correctly complete the following sentences:

A Recent developments in RTP have taken place because

Phrases

(*i*) the previous approach used in phosphorescence of freezing samples in liquid nitrogen is inconvenient in analysis.

(*ii*) it avoids problems due to scattering from turbid samples.

(*iii*) analytes can be examined in situ on filters and paper chromatograms.

(*iv*) the use of pulsed sources and electronic gating has made phosphorescence measurement easier.

(*v*) it provides a method of analysis for samples which do not fluoresce.

B Solid samples can be successfully examined by RTP because

Phrases

(*i*) the technique is not affected by particle size.

(*ii*) scattering is excluded by observing the emission when the sample is not being irradiated. ⟶

SAQ 5.4a (cont.)

(*iii*) quenching by collision is prevented due to the lack of mobility in the solid phase.

(*iv*) frontal illumination can be used.

(*v*) no sample preparation is required.

Response

(*A*) (*i*) and (*iv*).

Advances in instrumentation have made phosphorescence a viable analytical technique but only if the complications of low temperature sampling can be removed. (*iii*) is a consequence of this development work not the reason for it. (*ii*) and (*v*) are not sufficient in themselves to stimulate the interest in RTP.

(*B*) (*ii*), (*iii*) and, possibly, (*iv*).

Scattering is a serious problem with fluorescence and uv absorption methods (and also ir to a lesser extent). The lack of mobility in the solid state makes RTP possible. Frontal illumination does play a part (*iv*) but only because scattered radiation does not interfere. Particle size is still a problem (*i*) and the fact that no sample preparation is required does not guarantee that the technique will be successful (*v*).

****

SAQ 5.4b

Fill in the blanks in the following paragraph using words or phrases selected from the list below:

Phosphorescence may be observed from fluid solutions at room temperature by taking steps to reduce the .... between analyte molecules and potential .... such as dissolved oxygen. This can be done by .... the viscosity of the solution by .... the temperature or adding a viscous compound such as .... Alternatively, the excited analyte molecule can be protected by absorption into the .... region of a .... formed by the .... of long-chain fatty acids, particularly in the presence of heavy cations such as .... which facilitate .... to the triplet state. Certain types of organic molecules such as .... have a similar structure and behave in the same way.

hydrophilic	lipophilic
increasing	lowering
inter-system crossing	quenching agents
rate of collision	ethanol
cyclodextrin	antigen
glycerol	anions
glucose	derivatising agent
potassium	micelle
thallium	

Response

Phosphorescence may be observed from fluid solutions at room temperature by taking steps to reduce the *rate of collision* between analyte molecules and potential *quenching agents* such as dissolved oxygen. This can be done by *increasing* the viscosity of the solution by *lowering* the temperature or adding a viscous compound such as *glycerol*. Alternatively, the excited analyte molecule can be protected by absorption into the *lipophilic* region of a *micelle* formed by the *anions* of long-chain fatty acids, particularly in the presence of heavy cations such as *thallium* which facilitate *inter-system crossing* to the triplet state. Certain types of organic molecules such as *cyclodextrin* have a similar structure and behave in the same way.

****

SAQ 5.5a

0.0453 g of papaveretum (an opium concentrate) was weighed out, dissolved in sulphuric acid (0.05 mol dm^{-3}) and made up to 50 cm^3. A 10 cm^3 aliquot of this solution was transferred to a separating funnel and shaken with dry, ethanol-free chloroform (15 cm^3) for one minute. When the phases had separated, the chloroform (lower) layer was discarded and the aqueous phase was washed quantitatively into a 50 cm^3 standard flask with distilled water and made up to the the mark. Separate 10 cm^3 aliquots of this solution were transferred to two 50 cm^3 standard flasks. One of these was made up to the mark with sulphuric acid (0.05 mol dm^{-3}) and labelled 'solution A'. The other was made up with sodium hydroxide (0.1 mol dm^{-3}) and labelled 'solution B'. After purging the solutions, the blanks and the 2-aminopyridine standard with oxygen-free nitrogen for 10 minutes,

⟶

SAQ 5.5a (cont.)

the following fluorescence readings were obtained with the excitation monochromator set to 285 nm and the emission monochromator to 345 nm:

Solution A	574
Solution B	112
Sulphuric acid (0.05 mol dm^{-3})	27
Sodium hydroxide (0.1 mol dm^{-3})	43
2-aminopyridine standard	924

(*i*) Calculate the percentage of codeine and morphine in the sample using the calibration curves in Fig. 5.5a.

(*ii*) Which is the least accurate value in this calculation?

(*iii*) Is three-figure accuracy justified?

(*iv*) What is the purpose of ratioing the readings on the samples to that for 2-aminopyridine in this method?

Response

(*i*) The correct answer is codeine = 4.27%

morphine = 47.7%

If you got figures close to these (allowing for the errors in reading the calibration graphs) ... Well done!

If your answers are significantly different let's go through the calculation step by step to see if we can find out where you went wrong.

First of all, we must convert the fluorescence readings to I_{AP} values based on a reading of 100 for the 2-aminopyridine standard. Unless you do this you will not be able to use the calibration graphs which use I_{AP} as the ordinate. The 2-aminopyridine standard gives a reading of 924 at the same sensitivity setting as the readings for solutions A and B and the blanks. Before we convert the readings to I_{AP} values we can save a bit of time by subtracting the blanks first – sulphuric acid (0.05 mol dm^{-3}) from solution A and sodium hydroxide (0.1 mol dm^{-3}) from solution B.

Blank-corrected fluorescence readings:

$$\text{Solution A} = 574 - 27 = 547$$
$$\text{Solution B} = 112 - 43 = 69$$

If you failed to do this, your results would have been high – particularly for codeine. If you mixed up the blanks, the answers would have been a little high for codeine and (perhaps) a little low for morphine though the effect on the higher reading for the acid solution is less severe.

Now convert the readings to I_{AP} values. For solution A this is

$$100 \times 547/924 = 59.2$$

(ie the fluorescence reading is expressed as a percentage of the 2-aminopyridine reading.)

Similarly, the I_{AP} reading for solution B is

$$100 \times 69/924 = 7.5$$

Now we can refer to the calibration graphs to convert these I_{AP} values to concentrations but we must be careful! There are four graphs and we use the right one in each case! Remember that the fluorescence of morphine is negligible in alkaline solution so we start with solution B and read off the codeine concentration directly from graph B.

$I_{AP} = 7.5$ for solution B so

the codeine concentration is 1.97 mg dm^{-3}.

Now we take solution A in which both alkaloids fluoresce. Since we know the concentration of codeine we can calculate the contribution it makes to the fluorescence from graph A.

1.97 mg dm^{-3} codeine in sulphuric acid gives an I_{AP} value of 8.2. We must subtract this value from the I_{AP} value for solution A to get the contribution from morphine.

$$\text{Morphine } I_{AP} = 59.2 - 8.2 = 51.0$$

This gives a value for the concentration of morphine from curve C of 16.66 mg dm^{-3}.

Since the concentration of morphine is much higher than that of codeine, its contribution to the fluorescence of solution B will *not* be negligible and we need to go back and make a correction to the codeine concentration.

From curve D, a concentration of 16.66 mg dm^{-3} of morphine will contribute to the I_{AP} value of solution B to the extent of 0.16. Hence the codeine contribution is $7.5 - 1.6 = 5.9$. Referring this value to curve B gives a corrected codeine concentration of 1.55 mg dm^{-3}.

Now we have to redetermine the morphine concentration with the revised value for codeine.

Contribution of 0.59 mg dm^{-3} of codeine to I_{AP} for solution A

$= 6.4$ from curve A.

Contribution of morphine to I_{AP} for solution A

$= 59.2 - 6.4 = 52.8.$

Concentration of morphine in solution A

$$= 17.30 \text{ mg dm}^{-3} \text{ from curve C.}$$

The corrected values for the concentrations of both alkaloids in solution A are therefore:

codeine = 1.55 mg dm^{-3} morphine = 17.30 mg dm^{-3}

We could repeat the correction with the new value for morphine but the difference in the fluorescence of solution B caused by this change in concentration is negligible and further 'iteration' as it is called (a standard procedure in computer processing programs) is not worth while. (Try it and see!)

We still have to relate the concentrations in the solutions we measured back to the original papaveretum sample – a factor which is often ignored by inexperienced students of analytical chemistry who are, perhaps, dazzled by the intricacies of the instrument they are using or the elegance of the method! You, of course, would realise that the person who brought in the original sample will expect to be given figures which apply to that sample and not to some solution made up from it! This involves considering the dissolution and dilution steps leading to solutions A and B and 'back-tracking' with the results on those solutions. It is at this stage that things are likely to go wrong in the calculation – the following results are my second attempt which may be some consolation to you!

Solution A contains 1.55 mg of codeine in 1000 cm^3. The total volume of solution A is only 50 cm^3 so it contains 0.0775 mg of codeine $(1.55 \times 50/1000)$.

Similarly the morphine content of solution A is 0.865 mg $(= 17.30 \times 50/1000)$. Note how both values are expressed to 3 significant figures.

These two quantities were originally present in the 10 cm^3 aliquot taken from the 50 cm^3 aqueous phase after chloroform extraction.

Hence:

Total codeine in aqueous phase = 0.0775 × 50/10 = 0.387 mg.

Total morphine in aqueous phase = 0.865 × 50/10 = 4.32 mg.

These quantities were present in the 10 cm^3 aliquot taken from the the sample solution for extraction. Hence the content of the entire 50 cm^3 of sample solution is

0.387 × 50/10 = 1.93 mg of codeine and
4.32 × 50/10 = 21.6 mg of morphine.

The sample weight dissolved in 50 cm^3 was 0.0453 g so the

codeine content = 100 × 1.93/45.3 = 4.27% and
morphine content = 100 × 21.6/45.3 = 47.7%

This is quite a long and involved (though not difficult) calculation and so, if you did not get the correct answer, you could have made a mistake at any stage. If you cannot locate your error from the solution as set out above you may have used a different method which may be equally correct if you had not made a slip somewhere. If so, go back through your calculation and check or calculate if necessary the total content of codeine and morphine in mg in:

(*a*) 50 cm^3 of solution A,
(*b*) 50 cm^3 of extracted aqueous phase and;
(*c*) 50 cm^3 of sample solution.

This should locate the section of the calculation where you went wrong. The mistake could be in the arithmetic or, more likely, in the logic (not allowing correctly for volumes taken for transfer or dilution). You may also have made a copying error from one line to the next or used the wrong calibration graph. A fruitful cause of error is careless or sloppy presentation. Always write out each step in full as in the above calculation which is intended to be a model for this type of working. Calculations consisting of numbers without linking words of explanation or identification almost invariably go wrong and are virtually impossible to check!

There will be small deviations due to errors in reading the calibration curves (yours and mine!) but our results should agree to within 2 figures at least. If you quote answers to more than 3 significant figures consider yourself reprimanded! It is most important not to quote results to a higher precision than that of the least precise experimental value – including values generated by taking differences during the calculation, a common cause of loss of precision.

(*ii*) The blank corrected fluorescence reading for solution B (69).

(*iii*) Three-figure accuracy is not justified for codeine since this concentration depends on the fluorescence reading for solution B. (You must be careful not to leave in any additional figures introduced when converting these readings into I_{AP} values. Your calculator will probably give you 8 figures at this stage! In general, it is a good idea to work to one more figure than the experimental accuracy justifies during intermediate stages in a calculation to avoid accumulating rounding-off errors at the end.)

The value for morphine *is* correct to 3 figures because it is based on a 3-figure value for the fluorescence and subtraction of the codeine contribution does not affect the uncertainty in the third figure.

(*iv*) The ratio is independent of any variation in the source output or detector response with time or on different days. If, for example, the source output falls both the standard and the sample readings fall to the same extent and the ratio remains the same.

****

SAQ 5.7a

Identify in the following list, the operations which have to be carried out precisely (ie to ±1% or better) during the progesterone immunoassay because they affect the precision of the final results:

(*i*) Weighing 156 mg of progesterone to make up the stock solution.

(*ii*) Weighing 5 mg of progesterone-3-carboxymethyloxime to prepare the labelled antigen.

(*iii*) Measuring the absorbance of the stock solution of labelled antigen.

(*iv*) Measuring the fluorescence of the incubated solutions.

(*v*) Measuring the volume of antiserum to be added to the assay buffer.

(*vi*) Dispensing 100 μl of standard to the incubation tube.

(*vii*) Measuring 2 ml of hexane with which to extract the serum.

(*viii*) Addition of 300 μl of assay buffer to the sample and standards.

Response

(*i*) (*iv*) (*vi*) (*vii*) (*viii*).

Making up (*i*) and measuring out (*vi*) standard solutions always requires the highest level of precision because this determines the accuracy of the calibration graph.

The volume of buffer added has to be precise (*viii*) to ensure that the quantity of antibody and labelled antigen added to each sample is kept constant. As we saw previously, the exact weight of antigen in the buffer is not critical (*v*) when a calibration method is used though it would be if the result were calculated from the equations.

The volume of hexane used to extract the serum must be the same for each sample because the quantity of progesterone assayed depends upon taking an exact aliquot portion (500 μl) in each case. Taking precisely 2 ml is the most practicable way of ensuring that the volume is constant (*vii*).

Clearly, measurement of the fluorescence intensity must be precise since the concentration is determined by it directly (*iv*). On the other hand, the quantity of labelled antigen added to the buffer need be only approximate provided the volume dispensed is constant. Thus a precise measurement of the absorbance is not necessary (*iii*).

The weight of progesterone derivative used in the labelling reaction, like reagents in all preparative procedures, is not required to analytical precision.

****

SAQ 5.7b

Seven standard solutions, prepared and measured in a fluoroimmunoassay of progesterone, gave the following fluorescence intensities:

Concentration in Serum (nmol l^{-1})	Fluorescence Intensity (arbitrary units)	
0	700	711
5	671	680
10	592	611
20	493	503
40	410	422
60	333	342
80	299	311
100	279	288

The background fluorescence from the reagents other than the progesterone–fluorescein complex is 36 arbitrary units. Plot the calibration graph based on these data.

What is the most striking feature of the shape of this graph? Estimate the dynamic range of the analysis.

Response

The graph is shown in Fig. 5.7d.

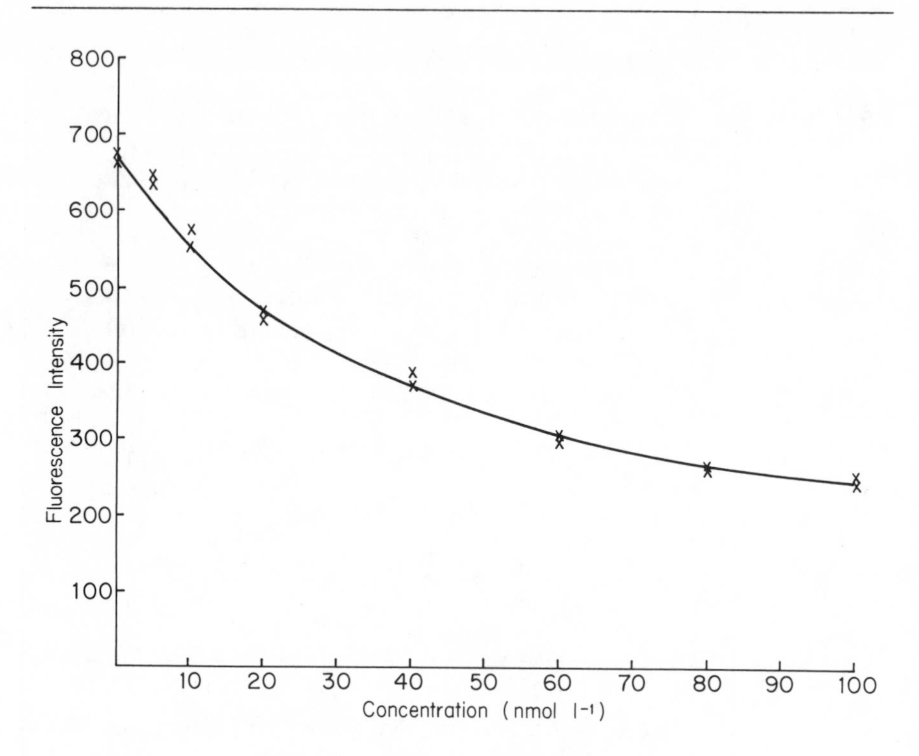

Fig. 5.7d. *Calibration graph for the fluoroimmunoassay of progesterone*

Both values are plotted for each solution – you may have taken a mean which is less satisfactory when the number of values is limited and the data not very precise. By plotting all the points, the general fit of the curve identifies values which are least reliable ('flyers'). Whichever method you adopted you should have subtracted 36 from all the readings (or means) to allow for background fluorescence. Note the downward curvature and the 'levelling off' at high concentration. This limits the dynamic range to about 50 (2–100 nmol l^{-1}).

****

SAQ 5.7c

The serum taken from four women patients at 3-day intervals over a period of 2 weeks was analysed by this procedure and gave the following results:

Patient	Day 1	Day 4	Day 7	Day 10	Day 13
A	655	650	548	450	410
B	641	628	637	644	635
C	445	472	642	654	650
D	295	313	335	356	368

Determine the concentration of progesterone in each sample using the calibration graph you have just drawn and give a clinical diagnosis of the condition of the four women (use Figs. 5.7a and 5.7b). In which case would further tests be desirable?

Response

The concentrations are shown (in nmol l^{-1}, the concentration unit of the calibration graph) in the following table:

Patient	Day 1	Day 4	Day 7	Day 10	Day 13
A	1.2	1.5	10.2	23.0	31.7
B	2.4	3.2	2.5	1.8	2.2
C	23.7	19.5	2.4	1.2	1.5
D	64.5	57.0	50.5	44.0	40.5

In order to refer these concentrations to Figs. 5.7a and 5.7b we have to convert them to ng ml^{-1}. The relative molar mass of progesterone is 312 (see 5.7.2) so

$$1 \text{ nmol } l^{-1} = 312 \text{ ng } l^{-1} = 0.312 \text{ ng ml}^{-1}$$

For the purposes of this exercise it is sufficient to divide the values in the table by 3 to get a concentration in ng ml^{-1} since it is not possible to read concentrations accurately from Figs 5.7a and 5.7b and we are only looking for trends anyway. From these graphs we may make the following observations:

Patient A

The values rise from a low 'background level' on Day 4 to about 10 ng ml^{-1} on Day 13 which corresponds to the concentration at the maximum on the 8th day after ovulation. Ovulation therefore must have occurred on or about Day 5 of these tests. The rise in the progesterone level at the start of pregnancy is not very different from the values obtained with this patient. However, by the 13th day of pregnancy the level would have reached about 13 ng ml^{-1} which, although some 30% higher than maximum in the menstrual cycle, is only just significant in view of the small scale of Fig. 5.7b. (Note that the origin of the time axis is essentially the same in both these graphs since conception (the start of pregnancy) can only occur within 2–3 days of ovulation. A further measurement on Day 16 would establish which curve was being followed – Fig. 5.7a if the value falls, Fig. 5.7b if the value rises. Alternatively a more specific pregnancy test could be carried out.

Patient B

This case is quite straightforward with values fluctuating around a mean of less than 1 ng ml^{-1} and shows that the patient is probably in the first half of the menstrual cycle. However subsequent measurements are desirable to determine whether ovulation does occur. The patient might be suffering from some defect leading to infertility.

Patient C

These values show the normal decline after Day 8 of the menstrual cycle for a healthy and fertile woman. The levels are too low to correspond to the dip in the pregnancy curve after Day 25 and too slow to correspond to the drop at the termination of pregnancy, (which in any case would be evident without having to carry out clinical tests!).

Patient D

These values are all well above the maximum in the menstrual cycle and fall away in line with the pregnancy curve following the maximum at Day 25 to the minimum at Day 40. The results appear to be a clear indication that the patient is pregnant but again a confirmatory test for pregnancy would be advisable in case the high progesterone level was due to some other cause. The rate of decrease in the progesterone concentration is too slow to correspond to the fall at the end of pregnancy after Day 270 which again would be an obvious condition.

Notice that the trend of the results is most important in clinical diagnosis of this type and that single measurements would be much less valuable, even if they could be made with high accuracy. Indeed, the fact that in 3 cases out of 4 further tests are desirable show that the results we have been considering are not really sufficient to establish the trend – and account for the lack of sensitivity of progesterone as an indicator in pregnancy testing.

****

SAQ 5.8a

RTP in both micelles and cyclodextrins requires the presence of a 'heavy atom'. Describe *three* ways in which this may be introduced, indicating in each case whether the method applies to micelles, cyclodextrins or both.

Response

(*i*) Use of thallium salts of fatty acids as micelles

(*ii*) Use of the technique for analytes containing bromine or similar elements.

(*iii*) Addition of a small molecule containing bromine or a similar element to the sample.

Obviously (*i*) applies to micelles only. (*ii*) applies to both though it limits the range of analytes which can be determined. (However, there is always the possibility of introducing a bromine atom for example into the analyte molecule by chemical treatment – a sort of derivatisation.) (*iii*) extends the range of analytes which are phosphorescent in cyclodextrins. It has not been reported with micelles but there would seem to be no reason why a similar procedure should not be equally successful.

****

SAQ 5.8b

Which of the following items constitute an advantage of cyclodextrins over micelles in RTP of fluid samples?

(*i*) Very wide application.
(*ii*) No need for de-aeration before measurement.
(*iii*) Longer lifetime of phosphorescence.
(*iv*) Requires the presence of a heavy atom.
(*v*) More selective.

Response

(*ii*), (*iii*) and (*v*).

Only those molecules small enough to fit into the cyclodextrin cavity show enhanced phosphorescence. This limits the range of applicability on the one hand (*i*) but provides a method of discriminating between analytes on the other (*v*). Oxygen quenching removes about 90% of the phosphorescence in cyclodextrins but the remaining 10% is sufficient to enable many analyses to be carried out without de-aeration (*ii*). Phosphorescence lifetimes in micelles are of the order of 1 μs which is too short for convenient time resolution so the longer lifetimes in cyclodextrins allows for less costly instrumentation (*iii*). Both micelle- and cyclodextrin-enhanced phosphorescence require the presence of a heavy atom (*iv*) so this is not an advantage for cyclodextrins over micelles.

****

SAQ 5.8c

Phosphorescence can be observed in the presence of scattered radiation either by the use of a total luminescence spectrometer of high performance or a fluorescence spectrometer with a pulsed source. Explain the principles of these two methods.

Response

In a total luminescence spectrometer, the phosphorescence is separated from the scattering (and the fluorescence) by its longer wavelength. A very low stray light specification is required.

With a pulsed source it is possible to use time resolution to take advantage of the longer lifetime of the phosphorescence, which is observed during the dark periods between pulses, by means of an electronic gating system.

Units of Measurement

For historic reasons a number of different units of measurement have evolved to express quantity of the same thing. In the 1960s, many international scientific bodies recommended the standardisation of names and symbols and the adoption universally of a coherent set of units—the SI units (Système Internationale d'Unités)—based on the definition of five basic units: metre (m); kilogram (kg); second (s); ampere (A); mole (mol); and candela (cd).

The earlier literature references and some of the older text books, naturally use the older units. Even now many practicing scientists have not adopted the SI unit as their working unit. It is therefore necessary to know of the older units and be able to interconvert with SI units.

In this series of texts SI units are used as standard practice. However in areas of activity where their use has not become general practice, eg biologically based laboratories, the earlier defined units are used. This is explained in the study guide to each unit.

Table 1 shows some symbols and abbreviations commonly used in analytical chemistry; Table 2 shows some of the alternative methods for expressing the values of physical quantities and the relationship to the value in SI units.

More details and definition of other units may be found in the *Manual of Symbols and Terminology for Physicochemical Quantities and Units*, Whiffen, 1979, Pergamon Press.

Table 1 *Symbols and Abbreviations Commonly used in Analytical Chemistry*

Å	Angstrom
A_r(X)	relative atomic mass of X
A	ampere
E or U	energy
G	Gibbs free energy (function)
H	enthalpy
J	joule
K	kelvin (273.15 + t °C)
K	equilibrium constant (with subscripts p, c, therm etc.)
K_a,K_b	acid and base ionisation constants
M_r(X)	relative molecular mass of X
N	newton (SI unit of force)
P	total pressure
s	standard deviation
T	temperature/K
V	volume
V	volt (J A^{-1} s^{-1})
a, a(A)	activity, activity of A
c	concentration/ mol dm^{-3}
e	electron
g	gramme
i	current
s	second
t	temperature / °C
bp	boiling point
fp	freezing point
mp	melting point
$\approx$	approximately equal to
$<$	less than
$>$	greater than
e, exp(x)	exponential of x
ln x	natural logarithm of x; ln x = 2.303 log x
log x	common logarithm of x to base 10

Table 2 *Alternative Methods of Expressing Various Physical Quantities*

1. **Mass (SI unit : kg)**

$$g = 10^{-3}\ kg$$
$$mg = 10^{-3}\ g = 10^{-6}\ kg$$
$$\mu g = 10^{-6}\ g = 10^{-9}\ kg$$

2. **Length (SI unit : m)**

$$cm = 10^{-2}\ m$$
$$Å = 10^{-10}\ m$$
$$nm = 10^{-9}\ m = 10Å$$
$$pm = 10^{-12}\ m = 10^{-2}\ Å$$

3. **Volume (SI unit : m^3)**

$$l = dm^3 = 10^{-3}\ m^3$$
$$ml = cm^3 = 10^{-6}\ m^3$$
$$\mu l = 10^{-3}\ cm^3$$

4. **Concentration (SI units : mol m^{-3})**

$$M = mol\ l^{-1} = mol\ dm^{-3} = 10^3\ mol\ m^{-3}$$
$$mg\ l^{-1} = \mu g\ cm^{-3} = ppm = 10^{-3}\ g\ dm^{-3}$$
$$\mu g\ g^{-1} = ppm = 10^{-6}\ g\ g^{-1}$$
$$ng\ cm^{-3} = 10^{-6}\ g\ dm^{-3}$$
$$ng\ dm^{-3} = pg\ cm^{-3}$$
$$pg\ g^{-1} = ppb = 10^{-12}\ g\ g^{-1}$$
$$mg\% = 10^{-2}\ g\ dm^{-3}$$
$$\mu g\% = 10^{-5}\ g\ dm^{-3}$$

5. **Pressure (SI unit : N m^{-2} = kg m^{-1} s^{-2})**

$$Pa = Nm^{-2}$$
$$atmos = 101\ 325\ N\ m^{-2}$$
$$bar = 10^5\ N\ m^{-2}$$
$$torr = mmHg = 133.322\ N\ m^{-2}$$

6. **Energy (SI unit : J = kg m^2 s^{-2})**

$$cal = 4.184\ J$$
$$erg = 10^{-7}\ J$$
$$eV = 1.602 \times 10^{-19}\ J$$

Table 3 *Prefixes for SI Units*

Fraction	Prefix	Symbol
10^{-1}	deci	d
10^{-2}	centi	c
10^{-3}	milli	m
10^{-6}	micro	μ
10^{-9}	nano	n
10^{-12}	pico	p
10^{-15}	femto	f
10^{-18}	atto	a

Multiple	Prefix	Symbol
10	deka	da
10^{2}	hecto	h
10^{3}	kilo	k
10^{6}	mega	M
10^{9}	giga	G
10^{12}	tera	T
10^{15}	peta	P
10^{18}	exa	E

Table 4 *Recommended Values of Physical Constants*

Physical constant	Symbol	Value
acceleration due to gravity	g	$9.81\ m\ s^{-2}$
Avogadro constant	N_A	$6.022\ 05 \times 10^{23}\ mol^{-1}$
Boltzmann constant	k	$1.380\ 66 \times 10^{-23}\ J\ K^{-1}$
charge to mass ratio	e/m	$1.758\ 796 \times 10^{11}\ C\ kg^{-1}$
electronic charge	e	$1.602\ 19 \times 10^{-19}\ C$
Faraday constant	F	$9.648\ 46 \times 10^{4}\ C\ mol^{-1}$
gas constant	R	$8.314\ J\ K^{-1}\ mol^{-1}$
'ice-point' temperature	T_{ice}	273.150 K exactly
molar volume of ideal gas (stp)	V_m	$2.241\ 38 \times 10^{-2}\ m^3\ mol^{-1}$
permittivity of a vacuum	ϵ_o	$8.854\ 188 \times 10^{-12}\ kg^{-1}\ m^{-3}\ s^4\ A^2\ (F\ m^{-1})$
Planck constant	h	$6.626\ 2 \times 10^{-34}\ J\ s$
standard atmosphere pressure	p	$101\ 325\ N\ m^{-2}$ exactly
atomic mass unit	m_u	$1.660\ 566 \times 10^{-27}\ kg$
speed of light in a vacuum	c	$2.997\ 925 \times 10^{8}\ m\ s^{-1}$